The Political Economy of Sustainable Energy

Energy, Climate and the Environment Series

Series Editor: David Elliott, Emeritus Professor of Technology, Open University, UK

Titles include:

David Elliott *(editor)*
NUCLEAR OR NOT?
Does Nuclear Power Have a Place in a Sustainable Future?

David Elliott *(editor)*
SUSTAINABLE ENERGY
Opportunities and Limitations

Horace Herring and Steve Sorrell *(editors)*
ENERGY EFFICIENCY AND SUSTAINABLE CONSUMPTION
The Rebound Effect

Matti Kojo and Tapio Litmanen *(editors)*
THE RENEWAL OF NUCLEAR POWER IN FINLAND

Antonio Marquina *(editor)*
GLOBAL WARMING AND CLIMATE CHANGE
Prospects and Policies in Asia and Europe

Catherine Mitchell
THE POLITICAL ECONOMY OF SUSTAINABLE ENERGY

Ivan Scrase and Gordon MacKerron *(editors)*
ENERGY FOR THE FUTURE
A New Agenda

Gill Seyfang
SUSTAINABLE CONSUMPTION, COMMUNITY ACTION
AND THE NEW ECONOMICS
Seeds of Change

Joseph Szarka
WIND POWER IN EUROPE
Politics, Business and Society

Energy, Climate and the Environment
Series Standing Order ISBN 978–0–230–00800–7 (hb) 978–0–230–22150–5 (pb)

You can receive future titles in this series as they are published by placing a standing order. Please contact your bookseller or, in case of difficulty, write to us at the address below with your name and address, the title of the series and the ISBN quoted above.

Customer Services Department, Macmillan Distribution Ltd, Houndmills, Basingstoke, Hampshire RG21 6XS, England

---

The Political Economy of Sustainable Energy

Catherine Mitchell
Professor of Sustainable Energy Policy, Exeter University, UK

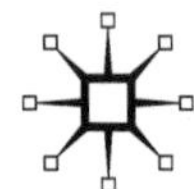

First published 2008
Published in paperback 2010 by
PALGRAVE MACMILLAN

Palgrave Macmillan in the UK is an imprint of Macmillan Publishers Limited, registered in England, company number 785998, of Houndmills, Basingstoke, Hampshire RG21 6XS.

Palgrave Macmillan in the US is a division of St Martin's Press LLC, 175 Fifth Avenue, New York, NY 10010.

Palgrave Macmillan is the global academic imprint of the above companies and has companies and representatives throughout the world.

Palgrave® and Macmillan® are registered trademarks in the United States, the United Kingdom, Europe and other countries

ISBN 978–0–230–53711–8 hardback
ISBN 978–0–230–24172–5 paperback

This book is printed on paper suitable for recycling and made from fully managed and sustained forest sources. Logging, pulping and manufacturing processes are expected to conform to the environmental regulations of the country of origin.

A catalogue record for this book is available from the British Library.

A catalog record for this book is available from the Library of Congress.

Printed and bound in Great Britain by
CPI Antony Rowe, Chippenham and Eastbourne

Contents

List of Abbreviations

AGR	Advanced Gas Reactor
BE	British Electric
BETTA	British Electricity Trading and Transmission Arrangements
BMU	German Ministry of Environment
BNFL	British Nuclear Fuels Limited
CAR	competitive at risk
CBI	Confederation of British Industry
CCGT	combined cycle gas turbine
CCS	carbon capture and storage
CERT	Carbon Emission Reduction Target
CNE	Spanish National Energy Commission
CHP	combined heat and power
CORWM	Committee on Radioactive Waste Management
CO_2	carbon dioxide
DEA	Danish Energy Authority
DEFRA	Department for Environment, Food and Rural Affairs
Dena	German Energy Agency
DERA	Danish Energy Regulatory Authority
DG	distributed generation
DNO	distribution network operator
DTI	Department of Trade and Industry (now Department for Business, Enterprise and Regulatory Reform)
DTI DGSEE	Department of Trade and Industry Centre for Distributed Generation and Sustainable Electrical Energy
EASC	Environmental Audit Select Committee
EC	European Commission
EDF	Electricité de France
EEC	Energy Efficiency Commitment
EECA	Energy Efficiency and Conservation Authority, New Zealand
EGWG	Embedded Generation Working Group
EU	European Union
EUETS	European Union Emissions Trading Scheme
EWP	Energy White Paper

FIT	feed-in tariff
GB	Great Britain
GBSO	Great Britain System Operator
GEMA	Gas and Electricity Markets Authority
GW	gigawatt
GWP	global warming potential
IEA	International Energy Agency
IFI	Innovation Funding Incentive
ILW	intermediate level wastes
IPCC	Intergovernmental Panel on Climate Change
kW	kilowatt
kWh	kilowatt-hour
LLW	low level wastes
LTS	large technical system
MED	Ministry of Economic Development, New Zealand
MfE	Ministry for the Environment, New Zealand
MP	Member of Parliament
mt	million tonnes
MW	megawatt
MWh	megawatt-hour
NAO	National Audit Office
NDA	Nuclear Decommissioning Authority
NETA	New Electricity Trading Arrangements
NFFO	Non-Fossil Fuel Obligation
NGO	non-governmental organization
NHS	National Health Service
NII	Nuclear Installations Inspectorate
Novem	Dutch Energy Agency
NZ	New Zealand
NZEECS	New Zealand Energy Efficiency and Conservation Strategy
NZES	New Zealand Energy Strategy
OECD	Organization for Economic Co-operation and Development
Ofgem	Office of Gas and Electricity Markets
PIU	Performance and Innovation Unit
POST	Parliamentary Office of Science and Technology
ppm	parts per million
PV	photovoltaics
PWR	Pressurized Water Reactor
RAB	Regulatory Asset base

RCEP	Royal Commission on Environmental Pollution
RCF	Rock Characterization Facility
R&D	research and development
RD&D	research, development and demonstration
REA	Renewable Energy Association
REC	Regional Electricity Company
RIA	regulatory impact assessment
RO	Renewables Obligation
ROC	Renewables Obligation Certificate
RPI-X	Retail Price Index minus X
RPZs	Registered Power Zones
RSP	regulatory state paradigm
SDC	Sustainable Development Commission
SE	Scottish Executive (now Scottish Government)
SME	small and medium enterprises
SRO	Scottish Renewables Obligation
TEC	transmission entry capacity
THORP	Thermal Oxide Reprocessing Plant
TNUoS	Transmission Network Operator Use of System
TO	Transmission Operator
UCL	University College London
UKERC	UK Energy Research Centre
WWF	World Wide Fund for nature

Acknowledgements

This book would not have come into being without the stimulation and help of a large number of people and institutions. There are those who have, over the years, caused me to get irritated enough with (sustainable) energy policy in the UK to want to write a book. There are numerous friends, colleagues and contacts I have had conversations and discussions with about (some of) the topics in the book. Others have sent me interesting papers and given me comments on chapters. Thanks to all of them. In addition, I would like to thank the ESRC for awarding me a Sustainable Technologies Programme Fellowship which enabled me to spend time thinking about sustainable energy in a somewhat broader way than I have done in the past and which gave me the time to write this book.

I would also like to thank David Elliott for being such a diligent and thoughtful series editor. Philippa Grand and Hazel Woodbridge at Palgrave Macmillan have been extraordinarily patient and helpful. Finally, I would like to thank Bridget Woodman, a colleague at Sussex, Warwick and Exeter Universities, who has been as patient and helpful as Philippa and Hazel while at the same time providing timely, insightful and practical comment and support throughout.

Series Editor's Preface

Concerns about the potential environmental, social and economic impacts of climate change have led to a major international debate over what could and should be done to reduce emissions of greenhouse gases, which are claimed to be the main cause. There is still a scientific debate over the likely scale of climate change, and the complex interactions between human activities and climate systems, but, in the words of no less than Governor Arnold Schwarzenegger, *'I say the debate is over. We know the science, we see the threat, and the time for action is now.'*

Whatever we now do, there will have to be a lot of social and economic adaptation to climate change – preparing for increased flooding and other climate related problems. However, the more fundamental response is to try to reduce or avoid the human activities that are seen as causing climate change. That means, primarily, trying to reduce or eliminate emission of greenhouse gasses from the combustion of fossil fuels in vehicles and power stations. Given that around 80 per cent of the energy used in the world at present comes from these sources, this will be a major technological, economic and political undertaking. It will involve reducing demand for energy (via lifestyle choice changes), producing and using whatever energy we still need more efficiently (getting more from less), and supplying the reduced amount of energy from non-fossil sources (basically switching over to renewables and/or nuclear power).

Each of these options opens up a range of social, economic and environmental issues. Industrial society and modern consumer cultures have been based on the ever-expanding use of fossil fuels, so the changes required will inevitably be challenging. Perhaps equally inevitable are disagreements and conflicts over the merits and demerits of the various options and in relation to strategies and policies for pursuing them. These conflicts and associated debates sometimes concern technical issues, but there are usually also underlying political and ideological commitments and agendas which shape, or at least colour, the ostensibly technical debates. In particular, at times, technical assertions can be used to buttress specific policy frameworks in ways which subsequently prove to be flawed.

The aim of this series is to provide texts which lay out the technical, environmental and political issues relating to the various proposed policies for responding to climate change. The focus is not primarily on the science of climate change, or on the technological detail, although there will be accounts of the state of the art, to aid assessment of the viability of the various options. However, the main focus is the policy conflicts over which strategy to pursue. The series adopts a critical approach and attempts to identify flaws in emerging policies, propositions and assertions. In particular, it seeks to illuminate counter-intuitive assessments, conclusions and new perspectives. The aim is not simply to map the debates, but to explore their structure, their underlying assumptions and their limitations. Texts are incisive and authoritative sources of critical analysis and commentary, indicating clearly the divergent views that have emerged and also identifying the shortcomings of these views. However the books do not simply provide an overview, they also offer policy prescriptions.

Nowhere is there more room for disagreement than in the debate over the best way to provide economic support for developing new energy technologies, and over how to choose which technologies to support. This opens up sensitive political issues, concerning the role of markets and their regulation.

This book argues that weakly regulated markets, based on narrow, short term economic considerations, are unlikely to deliver the technical and industrial innovations required to respond effectively to climate change: government must take a wider, longer term socio-economic view. This is a radical thesis, challenging the basis of much contemporary market-orientated policy in the energy field and beyond. However it may well be that if we are to deal with climate change, we will need to radically rethink our approach to economics and regulation, and adopt a far ranging revision of how energy policy is conceived, developed and followed up.

David Elliott

1
Breaking Free of the Band of Iron

Over the last few years, in response to climate change, the UK Government has produced a range of policies to stimulate the development of sustainable energy technologies (UCL, 2007; Mitchell and Woodman, 2004). These policies have been founded on market-based prescriptions coupled, where there are obvious market failures, with regulatory mechanisms. Together, these policies seek to use competitive pressures to drive specific areas of technology ahead. The overall approach is based on a set of principles and assumptions about how regulation and markets can, or should, work together and reflect the character of the underlying political-economic paradigm, which has been labelled the Regulatory State Paradigm (RSP) (Moran, 2003). Put simply, this paradigm suggests that Governments should provide a regulatory framework which 'steers' towards a defined general direction and then leaves it to the market to select the means to reach that end, although with some regulatory limitations.

However, as this book explains, this approach is unlikely to be sufficient given the need to radically redirect the economy in order to respond to the threat of climate change. There is a danger of ideological 'lock in'. A political paradigm establishes its own institutions and those institutions initiate policies based on the principles of the paradigm – currently, reliance on market competition as the main arbiter of value (Williamson, 2000). Those principles and policies promote narrow, short-term, economic considerations which are unlikely to deliver the technical, industrial, institutional and human innovations required. Government must intervene more effectively while taking a wider socio-economic view, to help stimulate development, deployment, acceptance, take-up and use of the relevant new technologies and associated infrastructure.

This requires either a new or an 'expanded' political paradigm, and a theme of this book is whether the latter would be sufficient. Nevertheless, the essential argument is that the current political paradigm is like a band of iron holding together a certain framework. Various actions can be undertaken around or between this framework but, in the end, this framework constrains certain actions or policies and defines the character of the paradigm. Until this 'band of iron' is broken, the UK can only do so much and no more in its quest to move to sustainable development.

This book focusses on the energy sector, but it is applicable to other sectors such as waste resources, agriculture and food policy or transport. It is an argument to move the basis of Government policy decisions (of all levels: local, regional, national and international) from narrow economic quantitative analyses to analyses which combine economic with technology and innovation theory; it is an argument to understand innovation from a systems perspective rather than from the current narrow technological perspective; it is an argument to move from the current under-valuing of qualitative social science to one which appreciates it and incorporates it in the policy framework. Together, these moves would add up to a political paradigm shift more able to deal with the climate change challenge. This book is intended to be one small step in that shift away from the current dominant political framework of the UK.

The challenge of climate change

In February 2007 the Intergovernmental Panel on Climate Change (IPCC) issued a report which concluded that the warming of the global climate system is now unequivocal. The planet has warmed by 0.74°C over the last century, and about 0.4°C of this has occurred since the 1970s. This warming is very likely to be because of the increase in anthropogenic greenhouse gas concentrations,[1] and the level of warming in future will be strongly dependent on emissions of greenhouse gases, including carbon dioxide. For a low emissions scenario, temperatures are projected to rise by 1.7°C, with a likely range of 1.1 to 2.9°C by 2090–99, in comparison with 1980–99. For a high emissions scenario, this increases to 4.0°C, with a likely range of 2.4 to 6.4°C (IPCC, 2007).

Furthermore, quite apart from 'dangerous' climate warming, the globe will warm in different amounts in different places leading to different effects. One major effect of this warming will be the increasing numbers of people displaced from where they currently live, with major implications for global unrest as they try to find somewhere to live and survive. Furthermore, most of the historical climate change emissions have arisen from the 'developed' world, while most of the earliest or worst effects are predicted to be in the developing world. Climate change is therefore an issue of enormous social and ethical dimensions. Despite this, the only international process to reduce carbon emissions – the Kyoto Protocol – is only in place until 2012. In planetary terms, meeting the Kyoto Protocol requirements is insufficient.

Overall, the UK has a target of reducing carbon dioxide emissions by 20 per cent from 1990 levels by 2010[2] and, ultimately, a 60 per cent reduction from 1990 levels by 2050 (DTI, 2003). The UK has also signed up to a binding target within the European Union of 20 per cent of total demand deriving from renewable energy in 2020; a non-binding reduction of 10 per cent of total energy; and a binding 10 per cent of transport fuels deriving from renewables. The burden share of this has not yet been agreed but it could be as high as 35 per cent of electricity coming from renewable electricity, in addition to significant amounts of renewable heat. Also, the UK has proposed setting binding targets for reducing carbon dioxide emissions through the Climate Change Bill 2007. These are a 60 per cent cut in carbon dioxide emissions from 1990 levels by 2050, and a 26–32 per cent cut by 2020. In addition, the Government proposes establishing five-year carbon budgets, beginning with the period 2008–12. Each budget will contain binding limits on CO_2 emissions.

As part of its strategy to achieve reductions in CO_2 emissions, the UK Government has set a number of targets for the increased use of renewable and combined heat and power (CHP) generation. These include: 'eligible' renewable generation to provide 15 per cent of electricity sales by 2015 and 20 per cent at an unspecified date in the future,[3] and the availability of 10 GWe of CHP capacity by 2010 (DTI, 2003, 2006a). There is also interest in the potential of micro-generation technologies, although the Government has so far not announced a target for capacity or output (DTI, 2006b).

So far, the UK is performing poorly against the Government's carbon dioxide target. The reductions that have been achieved were mainly due to the 'dash for gas' as gas-fired generation replaced coal in the early to mid 1990s. Although carbon dioxide emissions consequently fell from 1990 to the late 1990s, they have risen since the election of the Labour Government in 1997 and now stand at only 5.3 per cent below 1990 levels, well short of expected progress towards its 20 per cent reduction target by 2010 (DTI, 2007a). Despite this poor performance, the most recent projections very optimistically put emissions in 2010 at around 16 per cent below 1990 levels, but rising after that date (DTI, 2007b). This reduction, and reductions in other emissions such as methane, should however mean that the UK does at least meet its Kyoto commitment to get greenhouse gas emissions overall down by 12.5 per cent by 2012.

But very little of that will have been due to positive policy interventions in terms of low carbon energy supply or demand management – and these figures ignore the rapidly increasing emissions from air transport. Performance on renewables is similarly unimpressive: renewables supplied only 77 per cent of the Renewables Obligation target in 2005/06.[4] This equates to around 4.4 per cent of UK electricity generation in 2006 (DTI, 2007a). Levels of CHP generation have stagnated (Ofgem, 2006a; DTI, 2006a).

Translating urgency into action

To the average person on the street, climate change is probably something they are hearing more about, often in the context of: it's been the wettest/coldest/warmest day/month/year on record since records began. However, even those who take climate change seriously find themselves presented with a number of difficulties. Climate change appears so overwhelming, what is the point of doing anything? Even if they think they should do something, what is it that they should do? Can they afford to make the changes? Is it worth their while doing anything at all, given the cost and bother of doing things 'differently', particularly since it's a global problem?

The challenges facing Government are, unfortunately, even more complex. They fall (arguably) into four key areas: those relating to technology and innovation issues; those relating to what the role and relationship of Government should be and its relationship to the principal actors and stakeholders; those relating to the fact that the energy sector is privatized and regulated; and those relating to human consumption and behaviour. However, the over-riding issue that the Government faces is the *urgency* of climate change. In a perfect world, the Government might encourage elegant policies and incentives to unfold at their own rate and which lead to least-cost outcomes accepted by all. However, the reality is that Government has to start delivering change immediately. The flora and fauna of the globe, not to mention its peoples and future generations, do not have the luxury of time to allow perfect policies to develop: Governments and individuals have to get on with making greenhouse gas reductions immediately.

Within this context of urgency, the technology and innovation issues centre around the following very complicated questions: How should the current energy system make the transition to an affordable, low carbon and secure system? How can large amounts of low carbon renewable energy technologies, some of which are at an early stage of development, and demand side measures, be integrated in the operation and development of a sustainable energy system? What is the role of innovation within this system transition and how can it be stimulated?

Secondly, what is the role of (and between) Government, and its relationship to regulators, markets, businesses and customers in enabling the transition to a sustainable energy economy in a carbon constrained world. Should the respective roles and relationships continue as they are or should they change, and if so, how? What is the role of Government in assessing policies, costs and impact on individuals and UK competitiveness? Is any cost acceptable?

Thirdly, the energy sector is privatized and regulated. Even if Governments knew what they wanted and what path to take, what are the market and regulatory incentive frameworks that will deliver secure, cost effective and sustainable power and energy systems in a privatized world? What are the alternative long-term development and replacement strategies of the UK's ageing infrastructure and should they be speeded up to aid technology development? And if so, how? What are the security implications of a changing energy system?

Finally, what are the implications of climate change for human consumption and behaviour? Can Government 'allow' humans to continue to consume in the same way as they have done for the last several decades, or should there be greater curbs on this consumption? If so, how should these curbs be implemented: by market means or direct intervention (regulation). Put simply, if human behaviour is thought to be economically rational and thought to, ultimately, follow price signals, then consumption should be relatively easy to reduce via price signals. However, if humans consume and behave in ways which do not fit with rational economic choices, then curbing energy consumption and changing behaviour is much more complex than is recognized by economic principles and requires a greater range of more sophisticated policies and regulations.

This book argues that the current political paradigm (which includes the economic management of the country) and its underlying political principles are not only unable to deal effectively with the four key challenges of climate change, but they are also unable to deal with its urgency. This is because, fundamentally, the current political-economic paradigm involves processes of change that are too incremental and slow; because it has a linear, technological view of innovation which is unable to stimulate the appropriate system innovation; because it does not find it easy to recognize the non-linear, economically irrational behaviour of humans and consumption; because the roles and relationships between the institutions and actors are too constrained by the principles and processes of the political-economic paradigm; and because the current privatized and regulated world of the energy sector has ultimate responsibility to private interests rather than deliver the social good, even if some companies do their best to fulfil corporate social responsibility.

The regulatory state paradigm (RSP)

The UK's current, dominant political-economic paradigm or framework[5] (the regulatory state paradigm) (Moran, 2003) acts as a fundamental barrier to the move to sustainable development. Political paradigms have principles, and they establish institutions according to those principles, which in turn are the basis of their policies (Williamson, 2000; North, 1990). Thus, a political paradigm is the combination of the political (MPs and so on; their political concerns, such as being re-elected; political operators and their frame of operations; and the cultural, political context of society); the principles; the institutions and the policies. In the UK, there are several institutions which work in the energy sector. For example: the environmental regulator, the Environment Agency; the Carbon Trust; the UK Energy Research Centre. However, the institution which implements the rules and incentives by which energy is bought and sold in the UK is Ofgem, and in this respect is the most influential of those energy institutions.

The body of regulators, established since the mid 1980s, is an important aspect of the RSP. The RSP's principles underpin the rules and incentives of economic regulation, and the processes of the regulatory institutions enact those rules and incentives. They are, at root, discouraging rather than encouraging of the necessary innovation required for a move towards a sustainable energy system. Moroever, this book argues that it is a complex, but active, choice of Government to enforce its defining essentially market-orientated principles. As a result, policies not only in the regulated sphere but also those emanating from other Government actions are designed to fit within these principles. Together, the policies emanating from the 'political' and 'institutional' strand of the paradigm form the 'output' of the paradigm. This book argues that the constraints of the political paradigm are the central reason why the UK has been so poor in delivering a sustainable energy system. While the current RSP is able to deal well with certain economic issues, it is increasingly unable to adequately deal with complex and fast-moving problems facing society, such as sustainable development.

The fostering and enabling of the vision of a sustainable society requires the balancing of economic, environmental and social goals while taking account of security. This includes both price and non-price issues as well as long-term goals. The current paradigm is unable to do this balancing for two reasons. Firstly, because within the paradigm, the economic goal has de facto dominance; and secondly, because the political paradigm does not believe in pro-active 'balancing' or choice other than by competitive means via markets. This 'balancing' is essentially a political process and the institutions and philosophical framework of the RSP, and economic regulation which comes out of it, are unable to adequately perform it.

It is in this area of non-quantifiable qualitative factors that the RSP fails most crucially because, even if it wished to, it does not have the tools to take account of the non-quantifiable selection environment on an equal footing to the quantifiable factors. It also shows the difficulty of valuing those often qualitative effects. A new political framework which accepts the importance and value of non-quantifiable factors and therefore, on occasions, policies which are not the economic option, is needed if innovation on the scale required is to occur.

Moreover, security (including energy security) and sustainability are increasingly the central underpinning issues of global society. Both have different characteristics from many other important issues, such as manufacturing capabilities or competitiveness, because they are underpinning requirements of society, in the same way that water, food, shelter (housing) are. Answers to both of them have to be found; it is not that one is more important than the other. The principles do not value the core requirements of sustainability and security since neither of them are only or primarily economic issues. The political paradigm therefore finds it difficult to implement incentives to nurture them, and this matters to society.

The practical outcome of the regulatory state paradigm

In real terms, this means that when any Government, roughly supportive of the RSP, is working its way through the labyrinthine processes which take place behind any policy enactment, decisions and choices are made which fit with the political paradigm rather than based on the evidence of what works from technology and innovation case studies. This embodies an attitude to how technology development and innovation occur. A defining RSP principle is that innovation should occur through competition, based on choice by price within markets. This is discussed in more detail in the next chapter. However, such a principle implies a number of further attitudes which act to channel technological development and innovation down a particular route. In particular, it implies a view that risk (without a clear definition) is 'good' and that focussed technology support is 'bad'. This leads to the avoidance, where possible, of 'picking winners', meaning that choosing a technology to support would be considered to be 'intervening' in the market, thereby undermining the incentives of competition. Moreover, it assumes that investment in new technology will take place at the 'appropriate' time, given that investment risk. In other words, risk is not a 'good' or 'bad' thing but an intrinsic part of the economic choice.

The net result of the RSP and its economic regulation is that the UK gas and electricity market rules and incentives; network rules and incentives; the renewable energy policy; certain energy efficiency policies and increasingly micro-generation policies have been designed as far as possible to fit in with the principles which underpin the RSP. The design of these policies has a direct effect on the outcomes of these policies; on their indirect effects and their efficacy – and that design came straight from the RSP.

The RSP was meant to be a framework to improve government and delivery and yet with respect to sustainability this has not been the case. Policies are either not working (the Renewables Obligation – Mitchell et al., 2006; Szarska and Bluhdorn, 2006; Carbon Trust, 2006; EC, 2005a) or not working as well as they could be (combined heat and power – Owen, 2006); they may be taking too long (changes to the distribution network (Woodman, 2007a) and transmission access incentives (DTI DGSEE, 2007a)); they may not be in place (policies for renewable heat). At best they may therefore be having a limited impact (UCL, 2007); and certainly they may be far less ambitious than they could be (the Energy Efficiency Commitment, or EEC; biofuels transport obligation – REA, 2007a); and they do not enfranchise or enable participation of enough stakeholders (the Renewables Obligation, EEC, micro-generation linked to energy suppliers). As climate change becomes more and more important and as evidence of how different policies are working becomes clearer and more widely reported, then the basis of those policies will increasingly invite challenge.

A sustainable energy system

The challenge of successfully achieving a transition to a sustainable energy system, in the context of the UK's privately owned energy industry, rests on the ability of policy makers (at all levels and in all positions) to encourage and enable the necessary changes or innovation at the energy system level; at a firm level; but also in the patterns of sustainable consumption and behaviour across society. In the light of this, it would appear that a great deal of innovation is going to be required across the energy system. However, innovation is not linear; is not predictable; and it does have side-effects (Berkhout, 2002; Smith et al., 2004; Stirling, 2006a, 2006b; Geels, 2004a, 2006; Jacobsson and Bergek, 2004; Shove and Walker, 2007). Government action should be focussed on establishing a selection environment which is conducive to innovation (and this desired 'innovation' would be defined) and to try to 'channel' innovation as far as possible in the 'right' direction.

This requires enabling new management processes within the regulated network companies, licensed suppliers and generators, enabling far greater choice for customers, and enabling technological change within the energy system. This in turn requires change on a number of fronts: via privatized companies which have to modify themselves to be best placed in a carbon constrained world; via markets, so that new sustainable generation technologies (and their characteristics) can be incorporated; via network regulation so that the gas, electricity, heat and possibly other networks develop to enable rather than constrain change; via the regulatory process and via the selection environment – the nexus of political, legal, technical, institutional and social factors which establishes the benign conditions for innovation, thereby impacting on technology choice (Foxon, 2003; Foxon et al., 2005; Davies, 1996). In addition, this requires linking in with the other energy sectors besides electricity, such as transport, and the other sectors needing to change such as waste resources, agriculture and food policy.

Thus, the 'right' direction in this sense is taken to be harnessing the ability to make all the changes necessary to get from our current, dirty and polluting energy system to a sustainable energy system which results not only in carbon emissions that are as low as possible but also other sustainable outcomes such as reductions of the other greenhouse gases or radioactive wastes from nuclear power. A sustainable energy system would therefore be based on a combination of renewable generating technologies, renewable transport fuels, renewable heat, demand reduction, and the efficient use and integration where possible of heat and power loads by 'smart' information technology. A sustainable energy system could be made up of many different generating technologies, including more on-site and domestic generation, demand-side management policies, a change from a sales to a service culture, new management processes, more customer choice, new network management and design, including all sorts of new control technologies for network, generation and demand customers of different sizes.

Not all innovation is good innovation

In the face of this requirement for huge amounts of innovation, there are two, opposing views, which, put very simply, can be boiled down to opposite 'sides' in the innovation debate.

One argument is that climate change is such a huge problem, that any tool to combat it is useful and positive. This fits with the principles of the RSP. Non-interventionist policies can be applied. The rules and incentives behind markets and network regulation can be technologically and fuel 'blind', meaning that they are designed so that they do not favour any particular technology or fuel. Once these rules and incentives are in place, their outcome or effect (as opposed to a mandated or 'picked' outcome) can unfold in its own time.

A subsidiary but equally important policy choice for Government within this view, is the question of whether a market approach is better than a regulated approach; what mix of them is appropriate; and when a regulated approach should be used. For example, patio heaters might be wonderful in an individual sense, but they are a source of increasing greenhouse gas emissions. The same goes for numerous other 'new' inventions and innovations which are hugely enjoyable to a large segment of society – like plasma TVs. Within the RSP, there is a general support for a move away from 'nanny state' interventions. In principle, the Government is uncomfortable with allowing or disallowing a new technology to be sold in Britain by regulation. The preference is that the environmental externality of the product should be appropriately internalized by a monetary value and individuals should then choose whether or not to buy it based on price. The preferable way forward is to get the economic incentives 'right' and then leave it up to the consumer to choose whether they are prepared to pay. There have been occasions when market failures are considered so great that a direct intervention has been made, for example the coming ban on incandescent light bulbs. Nevertheless, the preference is that these types of interventions are minimized.

In principle, this argument is pursuasive. The problem arises in whether it is possible to get the incentives – the value of the externalities or the extra cost – right. As with incandescent light bulbs, the view was that this would not occur. Getting the 'right' cost is both an economic and a political question. Taxing a product, especially one so liked by broad swathes of society, is politically demanding. Moreover, it may be that despite the 'right' very high taxes, patio heaters may still be bought, and used, in large numbers. This may be the 'right' answer economically but not environmentally. And then finally, even if it is agreed that this is the preferred way forward, the 'right' value of the externality may not be reached; moreover, there may be other market failures or irrational (in economic terms) behaviour by consumers which leads to an economically inefficient outcome, i.e. more patio heaters being bought than should be. However, the latter may be entirely rational from the consumer perspective: they really enjoy patio heaters and they are going to buy one whatever the price, more or less.

The alternative argument to the 'all tools are useful approach' for combating climate change is that while a huge amount of 'innovation' is required urgently, unfortunately, not all innovation is 'good' or 'complementary'. Because of this, the 'everything is helpful' approach only makes the challenges of climate change worse (Mitchell and Woodman, 2006; Stirling, 2005, 2006a). This view to innovation requires Governments to make a choice: they have to attempt to follow a path. This is because different technologies require different commitments from Governments. For example, some technologies are more expensive and would require more Government resources; others are longer lived and therefore increase the risk of 'stranded assets' (or not being used until the end of their economic life). More contentious commitments might be for technologies which require greater demands on, or intervention in, the market; or when Government support for a technology maintains the status quo or momentum of the conventional system, thereby making it harder for new technologies to develop. It is possible that the practical outcome of a commitment by a Government to a technology, for example nuclear power, might have a deleterious effect on the development of other technologies (Mitchell and Woodman, 2006). In this situation, therefore, Governments cannot 'leave' it to technology and fuel blind markets and networks but do have to make a choice about what kind of energy future they want. In other words, they cannot follow the 'everything is useful' policy of combating climate change, because to do so will undermine other technologies.

Thus, not only does sustainable development require enormous amounts of innovation, it also requires the 'right' type of innovation. Moreover, because innovation is not linear, Governments should endeavour to establish a benign environment for innovation and to an extent 'shape' or 'channel' a move towards the 'right' direction.

At root, this requires Governments attempting to 'shape' or 'channel' innovation, even when they cannot be sure that it will be successful. Both these requirements – 'shaping' meaning directive choice, and doing this when they cannot be sure of the outcome – are not complementary to RSP preferences. Under the RSP, all the change and innovation required should derive from the rules and incentives derived from the principles of the RSP, the primary one being choice through the markets and competition. This reflects a view of 'innovation' as being linear and predictable; and that all outcomes or innovation are 'good'.

The innovation fault-line – the need for innovation or not?

One view of the move to a sustainable energy system is that it is essentially a technology question rather than a system question. If the transition to a sustainable energy system can occur by implementing new technologies within the current energy system, then very little has to change. The energy system can continue with more or less the same actors, except those actors will undertake their roles with some different technologies.

Put simply and as discussed in the next chapter, the regulatory state paradigm view of innovation:

- is undefined – all 'innovation' being good;
- supports economically rational (i.e. least cost) policies which complement large-scale, status quo companies rather than policies which encourage, create or reach multi-scale, multi-diverse unknown outcomes;
- believes in linear, predictable development or innovation (which enables a predicted known outcome from policies);
- considers quantitative economic analyses of markets, innovation and technological development as superior to broad qualitative analyses, not least because it finds the latter difficult to incorporate;
- considers broad carbon reduction policies superior to focussed, technology policies because the latter have to 'pick' a set of technologies or a particular technology;
- considers risk as an important stimulator in innovation while policies which reduce risk, inevitably, soften competitiveness which in itself must be undermining to incentives which lead to the 'right' answer.

As a result of these views, which derive from the principles of the paradigm, certain policies are put in place. This book argues in later chapters that if the Government of the UK (or of any other country) wishes to be successful in moving to a sustainable energy system then it has to move to the opposite side of the innovation fault-line, and this requires:

- an understanding of what 'innovation' is and that not all of it is 'good';
- an acceptance that markets are not the best way forward for making *all* choices – although certainly they are for many decisions (if not the majority) and will continue to be central to any future sustainable energy system;
- that it is not only acceptable to 'pick' a technology to support but necessary to 'channel' innovation policies;
- that choosing to support an environmental option, which may not be a least-cost measure, rather than choosing the economic or market option, may be appropriate, necessary and sensible and provide a great deal of additional value, albeit not in a way which is able to be valued monetarily;
- accepting that trying to meet the challenges of climate change is a 'system' issue not a technological-only issue.

The difficulty of making anything happen

The difficulty for Governments is that they are having to govern amidst competing wishes of society. While trying to balance all these wishes, they also have to be able to pick up key societal or global issues and 'lead' on them, often in the face of intense lobbying by sectors that don't want change – and incumbents rarely want change. Moreover, Governments want to stay in power and their timeframes are short.

As we have seen above, a great deal of 'good' innovation is going to be required. At the moment, this innovation is expected to occur as a result of appropriate and sophisticated signals through market and network rules and incentives. There are some mechanisms in place which soften this approach. Nevertheless, the overall thrust is still through this fundamental market approach.

Delivering change is in itself incredibly hard, given the process of government, and this is discussed in more detail in Chapter 3. As is shown, the process of government is very slow. The normal procedure on any given governmental issue is to have a Green Paper or draft policy, which is consulted on. Some of the points raised in the consultation will possibly be incorporated and become a White Paper – which becomes the formal policy, but which still has to be negotiated between departments and Ministers before it is put into practice. Very few policies get through which are in any way contentious. It is far easier for Governments to do nothing than it is to make change. And it's far easier for that change to occur within the political paradigm than for it to go against it. The Prime Minister is not immune from this process. At different times, he or she will have more power or more bargaining chips. He or she will not waste those moments of power but will use them for the policies they really want to get through. Climate change has had a good outing by this Labour Government, but it was not the issue on which Tony Blair was prepared to risk his bargaining chips. The same goes for Ministers. It is also true of large institutions, such as the energy regulator Ofgem. They also have a process of change which is slow and lends itself to outcomes equivalent to the lowest common denominator. And then, in parallel, despite it being hard to actually make a change, when policy change is under way, a process is kicked off which leads to other changes, some more important than others.

Is the UK Government starting to question its principles?

The UK has been very vocal about climate change. The Government has used its position as President of the EU and host of the G8 to argue for a more co-ordinated global response to climate change. In addition to this Government action, many other bodies and institutions in the UK, for example the Green Alliance and Institute for Public Policy Research, have also been extremely vocal in their efforts to press Government for even greater action. However, while vocal about climate change there is a powerful rift between different factions of Government about the basis of energy policy. This is a rift between those on different sides of the innovation fault-line.

The Labour Party came to power in 1997 and set about reviewing energy and environmental policies. It instituted the Utilities Act in 2000, which was effectively a re-regulation of the electricity and gas industries; it instituted an Energy Review by the Strategy Unit of the Cabinet Office in 2001–02 (PIU, 2002) which was effectively a Green Paper to the consequent Energy White Paper (EWP) in February 2003 (DTI, 2003). Since then it published another Green Paper in 2006 (DTI, 2006a); with the final Energy White Paper published in 2007 (DTI, 2007b). The two White Papers of 2003 and 2007 are markedly different. The question is which of the EWPs was the aberration? The former or the latter?

This U-turn question centres on three issues of concern: whether the market can be trusted to deliver; whether the policies that are in place are the right ones to meet the four goals of energy policy: security, social concerns, the environment and competitiveness; and whether the policies which derive from the political paradigm are acceptable to the public but also politicians. There is far greater transparency and evidence around the globe about the effectiveness of different sustainable energy policies, and this makes life harder for all Governments. The UK, which in many ways has been very vocal and forceful about the importance of climate change relative to most other countries, is doing badly in many keys areas, such as renewable energy deployment or demand reduction, compared to other countries which have not been nearly so vocal, for example Germany or Spain. This is particularly acute since so many credible commentators and reports have shown how poor the UK policies are (Carbon Trust, 2006; Rickerson and Crace, 2007); or have expressed preferences for non-market and non-competitive measures such as a feed-in tariff which guarantees a set payment for each kilowatt-hour (kWh) of renewable electricity presented to the grid operator (Stern Review); and which argue that the Renewables Obligation is a more expensive and less effective mechanism than the feed-in mechanism (EC, 2005a). Those in Government are aware of this and raise the pertinent questions: Why do we do what we do? And why are we so poor at delivering results?

And a further issue of whether the UK is following the right policies is the value of the outcomes from non-competitive policies. The UK has gone down the route of least-cost policies based on competitive and market policies. In theory, this should lead to the cheapest tonne of carbon or the cheapest renewable energy technology to be developed. It has benefited the large, now privatized but ex-monopoly companies, which have access to cheaper corporate finance and a greater ability to bear the (amongst other investment) risks, while it has not promoted new entrants. Non-competitive policies tend to have other impacts such as the stimulation of diversity (whether in terms of the size of power plants – for example, they tend to be a range of sizes; their type of investor – again they tend to encourage a broad range of investors from individuals through to large companies; their geographical position – they tend to use a broad range of resources, for example wind speeds; and they tend to promote a range of technologies). From the innovation perspective, new entrants may produce more innovation than incumbents (which are the older companies already in the energy system). Given that the cost of a competitive mechanism versus a non-competitive mechanism is more or less the same or even more expensive (EC, 2005a; Carbon Trust, 2006), questions, rightly, have been raised about why it is that we don't follow non-competitive measures since their indirect side-effects of diversity, for the same price, seem beneficial. Moreover, from the perspective of the RSP it seems 'wrong' that a competitive mechanism is not so successful as non-competitive mechanisms. This questions the basis of the RSP.

Moreover, another section of Government is pessimistic that the policies are 'right', in the sense it doubts that they will work. This group questions whether, irrespective of the political problems posed, it is possible to get individuals to do things they don't really want to: such as use less energy, have smaller cars, travel by air less, not use patio heaters or plasma TVs, and so on. This group questions the possibility of demand reduction to make any real demand reductions and whether large quantities of renewables are acceptable to the general public and will get through planning permission. This is a section of Government which looks backwards to state intervention in big supply projects, in this instance large nuclear power projects, so that low carbon electricity is available and individuals are able to carry on their lives more or less as they wish to. This view would see climate change as a technology issue, not a system issue; nothing really has to change in the energy system, there simply has to be more nuclear power.

Differences between countries

The earliest sustainable energy policies were put in place in the early 1970s (by Denmark and America) but the majority of European policies have been in place since 1990. It is now possible to understand why one policy worked and why another didn't, because the evidence and data are available.

This book argues that the political paradigm of a country has to essentially be on the pro-innovation side of the fault-line if it is to be successful in implementing and delivering sustainable outcomes. The country has to follow the innovation theory that high risk is not a good stimulator of technology development or investment; that the selection environment matters, and it is unwise to ignore it just because it seems complicated and is not easily quantifiable; and that there should be technology specific policies to help those technologies to develop through the maze of barriers, including those erected by momentum of the status quo energy system.

As the next chapter describes, the RSP has grown stronger and ever more inter-linked since the 1970s, and in particular since the arrival of Margaret Thatcher as Prime Minister in 1979. It has developed a momentum of its own, so that even though there are those that disagree with it, it will continue until there is a powerful movement for change.

A political paradigm, inherently, cannot be 'changed' at will. It reflects the consciousness of society. A paradigm is propelled into being by the very force which builds up behind it, and is then 'lodged' and codified through principles, institutions and policies. It remains there until the force of a new paradigm is built up, like the stretching of an elastic band, and propelled forward, knocking the old paradigm out of the way.

For UK energy policy, most of these questions are much debated. However, at the core of policy shaping, where principles and the essential design attributes are agreed, being on this side of the fault-line would still be unacceptable. At the moment, these debates are centred on the future of Ofgem, the energy regulator. The Sustainable Development Commission has undertaken a review of Ofgem and its impact on sustainable development (SDC, 2007). The Commission for Economic Markets and Environmental Performance, set up by the DTI, Defra and Treasury, is tangentially reviewing Ofgem's duties.

Governments are not monolithic. Not all parts of Government support the same thing. Different departments, local authorities, companies and individuals can put in place different policies. However, the essential argument of this book is that the political paradigm at its centre is like a band of iron holding together a certain framework. Certain actions can be undertaken around or between this framework but, in the end, this framework constrains certain actions and defines the character of the paradigm. Until this 'band of iron' is broken, the UK can only do so much and no more in its quest to move to sustainable development.

While the fundamental costs and revenues related to the rules and incentives of markets and networks regulated by Ofgem remain similar, even if Ofgem's duties are altered, the band of iron will remain intact. So far, while there have been skirmishes around that band of iron, nothing has so far dented it. Only when there is a direct decision to break that band will the costs and revenues of the actors in the energy system alter, thereby enabling new ways of doing things.

The layout of the book

The rest of the book explains this in more detail. Chapter 2 describes the regulatory state paradigm and shows why it is no longer able to answer the challenges of society. It sets out the key principles upon which RSP rests and then describes the challenges facing the RSP and why, inherently, the paradigm is unable to answer those challenges satisfactorily. It argues that the current model of regulation keeps the 'old' system going and that the downside of this is that it's great for the incumbents; but bad for consumers; bad for new entrants; bad for 'good' innovation; bad for Sustainability, with a capital S (Stirling, 2006b); and bad for global partnership.

Chapter 3 then discusses how difficult it is to make policy happen. It examines this in three ways: it sets out an argument of what a sustainable energy system is and how much change will be required; it then examines the process of government and the regulator and shows how difficult it is to make anything happen, never mind the huge changes required for the transition to a sustainable energy system; it then examines the innovation literature which sets out different arguments about how innovation happens. Within these three strands, the chapter looks at the situation of incumbents (the large ex-monopoly companies) and new entrants, and the importance of diversity (including scale, of technologies, resources, investors and customers).

This argument is amplified in Chapter 4 by arguing that the Government, in keeping with the paradigm, is taking the UK down the nuclear route and away from renewables and demand reduction. It describes the developing arguments for a new nuclear power programme in the UK. Nuclear power's strengthening re-emergence represents the Government finding policies to support its paradigm rather than appropriate evidence-based leadership. Support for nuclear power is a step back to 'club' politics but at the same time a continuation of the view that the large companies are really the key to getting us through the difficulties of climate change. However, this book argues that going back no more offers the answer to society's difficulty than did Keynesian Intervention, the last paradigm, when it was superseded by the RSP. Any new paradigm has to 'fit' the issues and concerns of today's society and has to resonate with the core concerns of that society for at least the next couple of decades. Support for monolithic large companies, manipulating the market in their favour and excluding new technologies and ideas, does not have that required resonance.

Chapter 5 illuminates the impact of the RSP principles on UK renewable energy policy. It concludes that the RSP has never taken renewable energy seriously, because it simply doesn't fit as a set of technologies to that paradigm. Chapter 6 then provides a case by case analysis of why the UK gas and electricity market rules and incentives and the network rules and incentives are as they are. This chapter explores the effects of the basic principles of the RSP in terms of sustainable energy delivery and policy success and its wider knock-on effects for climate change policy and sustainable development.

Chapter 7 undertakes a case study of New Zealand, a country which has a similar political paradigm to the UK albeit with very particular New Zealand charactersitics. It shows how the New Zealand political paradigm has undermined the development of climate change policies, despite the country's widely cherished 'clean and green' image.

Chapter 8 then examines Denmark, Germany, Spain and the Netherlands and asks whether they have been better able to deliver a sustainable energy system than the UK because of their different political paradigms. All are different in some way, although Germany and Spain are more similar than the other two. Denmark has followed a very individual path. As a country, its current domestic market is poor but it is still dominant in the world market for wind energy technology issues. The Netherlands is a particularly interesting case study because, unlike any other country, it has incorporated transition management policies into its sustainable energy policies. Despite this overt recognition of the importance of innovation, sustainable energy delivery remains poor.

Chapter 9 then pulls everything together. It examines the basis of countries which 'just do it'; that do seem to be on the 'right' side of the innovation fault-line; which do seem to be able to enable participation by their citizens; and which do seem to be able to deliver sustainable energy policies which work. It argues that in matters of climate change, the environmental choice should, in some situations, take precedent. This should be a political decision and be implemented through clear, legislated action. This would deliver the necessary powerful, interwoven framework needed which links policies, innovation, economic regulation, planning, consumption and technology issues to move to a sustainable energy economy. The chapter argues that this will require intervention in support of these sustainable technologies within markets and within economic regulation but that this does not preclude dynamic, competitive markets. It should be viewed as a necessary reduction in hurdles to enable the system-wide transformational forces to take place. Economic regulation would continue, but would be secondary to it in certain defined areas, but effectively in matters related to climate change. There are many examples of this in Europe.

The push for this has to come from the Government. Only the Government, through its determination and legislation, can provide confidence throughout the energy system and thereby stimulate the necessary investment and participation from all quarters. Other countries do this in parallel to successful economies. It is time that the UK started to do the same. This is not to argue that economic theory and competitiveness be sidelined – but that innovation and the importance of sustainability take their rightful place beside them.

Provisos and apologies

This book balances four themes: sustainable energy, political paradigms, sustainable development, and technology and innovation policy. I feel confident that I am knowledgeable in one of those four themes – sustainable energy – and passably informed in two others: sustainable development, and technology and innovation policy. The other area, that of political economy, has opened itself up to me as I have tried to understand why it is that certain Governments are so much more successful than others in combating climate change and in protecting the environment. In order to answer that question, it has been necessary to enter academic areas in which I am not specialized. I have, for example, more or less paraphrased the debate in the sustainable development literature about economic versus environmental balancing in little more than a paragraph; and the same goes for many other important areas of discussion, for example, risk and technology development. I believe the strength of this book is that it tries to link a number of areas. I am also aware, in a yin and yang sense, that this may also be its weakness. I hope that readers view the arguments in the book with good will and as a kick-off point of the debate.

Finally, I talk of the principles and policies of the RSP. I recognize that there is within that paradigm a spectrum of views, some of which will be utterly opposed to certain Government policies. This spectrum is discussed in Chapter 3. Every time I use the phrase 'political paradigm' I could reference it with a proviso, but the book wouldn't read well if I did. I apologize for the sometimes blanket statements. I have attempted to balance this by discussing in detail the spectrum of views within the paradigm and how that translates into policy, for example with respect to renewable energy policy development. Even so, I recognize that the widespread use of 'RSP principles' or political paradigm is achingly broad.

2
The Regulatory State Paradigm – and Its Challenges

In the UK, the relationship between state involvement and economic regulation has taken a number of twists and turns over the last century or so. Paraphrasing greatly, it has moved from pre Second World War 'club' government (the cosy management of Britain by friends and aquaintances) to Keynesian Interventionism (where the state intervened on a grand scale to improve the standard of living and energy supply for the mass of the British populace) through to today's political paradigm of the regulatory state (where the divide between public and private is based on legal distance and state institutional decisions are made, in theory, as a result of technocratic expert opinion and economic analyses).

There have been, and are, different types of state economic intervention. The main strands are the economic management of trade and commerce (i.e. managerial) and state intervention in big public projects, such as council house building or Concorde (i.e. investment). Both of these, led to either public ownership or management of public bodies.

The Conservative Party under Mrs Thatcher did away with public ownership, where possible, state management or intervention and changed the basis of what was called the 'welfare state'. The Labour Party under Tony Blair continued this approach, including new measures such as making the Bank of England independent (Bank of England, 2006). Markets are seen as the best way to drive the economy, with the state (allegedly) 'steering', as far as possible by setting the conditions for economic efficiency without old-style managerial intervention, and certainly without old-style public investment. As the 2007 Energy White Paper makes clear:

> A market-based approach within a clear policy framework provides an effective way to help us manage this uncertainty and deliver our policy goals. This is because companies are best placed to weigh up and manage the complex range of interrelated factors affecting the economics of energy investments.
>
> The private sector will be best able to help us deliver our goals and manage the associated risks when they have access to a wide range of low carbon investment options. The Government's role is therefore to provide a policy framework that encourages the development of a wide range of low carbon technologies, so we can minimise the costs and risks to the economy of achieving our goals. (DTI, 2007b)

This chapter endeavours to illuminate the character of the regulatory state paradigm; the principles it works to and which it has installed in the institutions it has set up; the implications of those principles; and the challenges which the paradigm now faces. From these sections, the intention is to reveal how unsuited the regulatory paradigm is for dealing with the complex challenges of climate change.

The basis of the regulatory state paradigm

The following section is indebted to the early chapters of Michael Moran's *The British Regulatory State* (Moran, 2003). The previous political paradigm – the Keynesian Interventionist State – came into being after the Second World War when there was a generalized move towards the view that the effort of the previous years should lead to a better standard of living and sense of inclusion for a wider proportion of the British population. This permeated society, but included the creation of the National Health Service; the mass building of council houses; and the opening up of state education, including at university level. There was a feeling that this could best be choreographed by the state and in this way, the three aspects of the left's political paradigms were brought together: the political (its operational machine – the MPs, its practical aspirations); its philosophy and principles, which derive from the wider progressive concerns about how society could and should develop, and upon which it rests and is defined; and its institutions, set up by (or altered to suit) the political framework, to deliver its vision and carry its principles in the form of policies (Moran, 2003; Williamson, 2000).

The gradual move from this approach of state intervention towards the current approach of the regulatory state paradigm developed from an increasing belief that UK society would be better off if there was a separation of Government and politicians from the management and ownership of industry, and with less intervention (and investment from public finances) in big public projects (Vickers and Yarrow, 1988; Baldwin and Cave, 1999). 'Club' government or the direct management of the Keynesian period gave way to a 'hands off' political framework, which included independent industry regulators who worked to legal duties. This paradigm was ushered in with the Thatcher Government of 1979, and the Keynesian Interventionist period of government was visibly well and truly over with the privatization of British Gas in 1984. Although the RSP came into being for a number of reasons, in essence it was because it 'fitted' with the broad ideas of what it was claimed society wanted for the future: free markets but with some regulation to prevent excesses, and a welfare safety net. UK society had got used to the benefits of state education, the health service and access to housing. But just as importantly, society was beginning to be affected by the inefficiencies of the UK's economy. The people of the UK wanted a more efficient management (Marr, 2007).

The essence of these RSP principles are that: markets and competition are seen as the most effective way of meeting society's choices; politicians should be legally separated from the regulation and decision-making of industry; the means of 'steering' the delivery of efficient management of the UK's industries should be based on 'expert' knowledge and economic analysis using open and transparent processes and data; markets should be designed to be technology and fuel blind so that outcomes are not 'picked'; if an outcome is wanted, the policy put in place should mimic markets as far as possible and should not intervene directly in the market or network rules and incentives (for example, the Renewables Obligation); as far as possible, direct regulatory measures should be instituted only in the face of substantial market failures (for example, the banning of incandescent light bulbs). It is these principles which inform and sometimes constrain, policies across Government, including (sustainable) energy policy. British Government, according to Moran, spent the first two thirds of the twentieth century 'slumbered in the historical equivalent of a long Sunday afternoon' (Moran, 2003 p3). It woke from this stagnation to 'hyper-innovation': the change which arose from the linking of two crises – a crisis in the content and outcome of economic policy in the UK; and a crisis in the system of government – 'club' government – itself. These crises led to a widespread appetite for fundamental policy and institutional change across society – and it is this combination of change which saw the transformation of the political paradigm and led the way in defining the boundaries between public and private, which were re-drawn in complex ways and had unexpected consequences.

At root, this pursuit of a new means of governance (politics and institutions), the re-definition of the boundaries of public and private, and the desire for competent economic policies, all made enormous sense when this confluence of change occurred. However, this book argues that this political paradigm is no longer appropriate for the carbon constrained world we find ourselves in and the pursuit of sustainability.

This book argues that it is the active choice of Government to enforce its defining principles; policies and economic regulation are designed to fit within these principles; and despite the UK's apparent very vocal desire to shift to a more sustainable energy system, it is failing to do so as a result of the misaligned principles of the RSP with the needs of sustainability. This is because the regulatory state paradigm is suited to dealing with economic (rather than non-economic) issues which can be dealt with by policies derived from quantitative (not qualitative) analyses. In addition, it is unable to implement the linked policies required of a successful innovation process to enable the transition from one energy system to another – it is unable to deal with the complexities of the challenge of climate change. Moreover, in its efforts to deal with the challenges of climate it necessarily installs policies which can deal with only part of the problem and which then act as further barriers; and its attempts to balance the economic, social, environmental and security goals of sustainable development and energy policy necessarily fail because of the de facto hegemony of the economic goal as a result of its legal duties. And finally, even if it could make these balancing decisions, regulators are inappropriate agents to undertake such balancing, since the latter are 'political' decisions.

When Moran termed this paradigm 'the British Regulatory State', he meant it in the sense of a regulator of a mechanical system: adjusting, reconfiguring and balancing social and economic systems. This is sometimes discussed in terms of 'steering' rather than 'rowing'. This is compared to the previous paradigm where Government both steered and rowed, meaning that they were heavily involved in the running of industry as well as deciding its direction. The new political paradigm wanted to establish the strategic direction of Government, rather than delivering services themselves. The 'regulatory state' is a confusing term for those who work in a world of privatization and liberalization. In that world, the word 'regulatory' has almost the opposite meaning as used by Moran, who uses it to describe the political paradigm championing liberalization and laissez-faire economics. To those who have not read his description and definition of the regulatory state, the word regulatory conjures up direct regulation and a continuation of command and control. Moreover, an important aspect of the political paradigm was the dismantling of regulation which restricted competition – so it is also de-regulation, as it's called in the US. This is confusing and as a result this book tends to talk of the new 'political paradigm', but when it does use the phrase 'regulatory state paradigm' or 'British regulatory state' it is using it in the Moran sense.

What the new paradigm stands for

Substantively, the new 'laissez-faire regulation' political paradigm turned to the reconstruction of institutions and economic practices in the UK with the aim of raising global competitiveness by increasing policy competence and economic efficiency and effectiveness. There were two main, inter-related means of doing this: firstly, by altering the way Government governed; and secondly, transforming the ambitions of Government by withdrawing from the interventionist projects which had accumulated over the last century, and by instituting regulators to regulate industries which had often previously been overseen by public bodies. The new political paradigm was the combination of a switch to new policies but also a switch to new institutions and attitudes. It wished to move away from 'club' government (Hood et al., 1999) – meaning its informality; reliance on the tacit knowledge of someone by virtue of them being an 'insider'; autonomy from public scrutiny and accountability; almost oligargic; secretive; with the marginalization of law. It wanted government to be more open and accountable in the technocratic sense and this therefore led to a paradigm with more public reporting; quantification; standardization; formality; and auditing which was intended to provide systematic information accessible to insiders and outsiders.

The 1970s was a time of fundamental change for the UK. Imperialism was a spent force, and an effect of this was that the legacy of Britain's pioneering industrial force, to an extent built on the resources available from the 'empire', was becoming exhausted. At the same time, the new forces of Europeanization and globalization appeared to have new requirements and possibilities. This fed forcefully into the re-creation of the institutions of government. Governing arrangements were to change so that codified knowledge would become more important than the tacit knowledge of those who 'knew'; similarly codified rules were to become more important than 'knowing' what was wanted; 'merit' or achievement was to become more important than traditional 'position'; and measurable accountability was to become more than elite solidarity.

Moran argues that this was Britain trying to become a 'modern' state. Moreover, it was

> trying to make transparent what was occluded; make explicit (and where possible measurable) what was implicit; and equip state with a capacity to a standardized view and to use this to pursue range of projects of different types of social control. (Scott, 1998)

The post-war Welfare State undertook ambitious, comprehensive intervention towards large-scale public ownership; large-scale direct social welfare provision; and purposeful economic management to fulfil employment. The new political paradigm responded by a systematic divestment of industries which had been publicly owned. Important aspects of this were the dismantling of regulation which restricted competition – and, as the term 'liberalization' implies, a fundamental reshaping of the way Government tasks were defined with respect to a public–private divide; a recognition of the limits to state resources; and a confining of itself to the creation of frameworks of rules for intervention. Regulatory agencies were created to oversee industries, thereby renouncing the command modes of the Keynesian era. Politicians were to be freed from interventionist (and necessarily politically one-sided) control in an effort to guide policy by technocratic and economic imperatives.

However, while the new political paradigm scaled down its ambitions of direct intervention, the regulatory mode greatly widened the range of economic and social life that was subject to public power. Arguably, its regulatory projects have been just as ambitious as the past Keynesian Interventionist model. So while it may have turned away from 'command and control' it has also led to an increase in growth of apparatus of control in the public sector: top down micro-management continued.

The basis of the vision behind the new political paradigm was to produce institutions (through policies and institutional reforms) and markets which would lead to more effective national success in a competitive world. Numerous policy actors from different ideological backgrounds have contributed to the reforms. The transition of the UK political framework to the new paradigm occurred when both the 'new' means of governance but also independent economic regulators became accepted as the normal means of governance for an industry. Moran argues, and this is discussed later, that the transformation to the new form of open and transparent, 'non-club' government did not fully happen. However, the institutional transition did occur and the regulators expanded in size and influence while politicians appeared distanced, as a result of legislation, from the day-to-day running of state companies or industries. This required establishing the idea that the regulator was 'independent', 'neutral' and 'value free' (unlike the politicians who would be partisan and short-term in their outlook). This 'make-over' partially occurred by setting out duties by which the regulator should work to. The duties were legislated through Parliament, placing legitimacy on the regulator but also clearly establishing its boundaries. Regulatory decisions were to be made on the basis of 'expert' technical and economic, rather than qualitative, analysis. This led to clear, unambiguous 'right' answers which allowed the development of the idea of 'credibility'. Because decisions were based on 'expert' technical knowledge and quantitative analyses, the decisions would be 'right' and could be relied on.

Thus, the crisis of 'club' government and the crisis of economic management of the country via Keynesian Interventionism led to the new political paradigm and also explains why the impacts of its implementation created such powerful ripples throughout society – it changed both the political landscape and the institutions and principles (and therefore applied policies) of Government. This led to a *new* kind of state which instituted a *new* kind of governance. This was not a retreat of Government (Scott, 1998), more a reshifting of tectonic plates which continued central control and the cascades of principles down to policy.

Originally, the positive side of the paradigm of injecting greater economic efficiency into the ex-monopoly companies and monolithic state organizations could be seen, albeit with major social effects which are not touched upon here. However, as time has gone on and the issues that the energy regulator(s) have had to deal with have altered in urgency or character, this inability to incorporate qualitative values can be seen as a central flaw.

Unintended consequences of regulation

As described, the new paradigm combines the political philosophies and principles and its institutions which enact those principles and philosophies. The key institutional change of the new paradigm was the establishment of economic regulators, across all sectors of society: hence Moran's term 'the British Regulatory State'. According to Moran, there have been four key unintended consequences of the new political paradigm, all of which are important for the pursuit of sustainability:

- A compromise on the basis of the new regulatory bodies.
- The widening of areas covered by economic regulation.

- A crisis of the benefits of the new political paradigm with the crisis of rail regulation in 2001.
- A creeping colonization of regulation into areas hitherto free of it, thereby creating a powerful web and 'corps' of regulators, and a momentum within that web.

The new political paradigm was a reaction against 'club' government and a desire for a regulatory world where tightly constrained non-discretionary technocratic judgements would maximize the natural workings of markets. This goal was fostered initially by a number of academics and intellectuals, such as Sir Keith Joseph, Michael Beesley and Stephen Littlechild. However, their vision of a democratic doctrine of accountability, openness and transparency was somewhat watered down by the time the regulatory legislation was put in place. This meant that the 'political' portion of the paradigm was able to act in ways more attuned to 'club' government when desired, despite the implementation of the institutional aspects of the new paradigm.

The second unintended effect was that the institutional strand of the new political paradigm has become asked to do what it was not set up, nor is able, to do. Privatization happened in the 1980s and early 1990s under the Conservatives. The newly elected and powerful Labour Party was swept into power in 1997. They set about what was essentially a re-regulation through the Electricity and Gas Acts in 2000, to include new aspects of economic regulation, such as social and environmental concerns, while re-jigging other aspects, such as increasing the importance of the interests of consumers. But social and environmental issues were never intended to be answered initially and have sat badly ever since.

The decision to try to include such concerns fits neither with Littlechild's vision of evidence-based, quantitative, openness and transparency *nor* with what was created via various pieces of legislation. The Labour Government wanted something more from the regulatory bodies: they have taken the fundamental thrust of the political paradigm which suited them but also added in other 'bits' more related to their own concerns (and the voters'), without any real discussion of how possible it was going to be to deliver what they said they wanted, given the principles and institutions in place. Because of this, the regulatory bodies have increasingly become part of a political process, despite continuous statements to the contrary by Government and the regulator, because of the pressure on them to fulfil the Government's environment and social policy. Regulators have a great deal of discretion when interpreting their duties. Arguably Ofgem has used that discretion in a very limited manner. Nevertheless, the regulator does work to these legal duties and Government cannot 'blame' the regulator when the regulator does not do what the Government wants. This is discussed further in Chapter 5.

A third unexpected consequence of the move to the new political paradigm has been its undermining from crises that have occurred directly as a result of its changes. In particular, the crisis of rail regulation in 2001 has led to major implications for wider regulatory goals and concerns. Moran argues that the significant damaging factor was that the crisis was essentially handled in a manner very similar to 'club' government, and this led to the widespread question of how much had governance really changed under the new paradigm.

- The regulators were swept aside and the Secretary of State was seen to be making decisions in a high profile manner. It thereby re-politicized the regulation debate, shifting the debate about regulation (control or incentivization) of industry up the chain back to Ministers – and this despite the fundamental aim to de-politicize.
- It put corporate form and purpose back on the open political agenda. This was because the Railtrack board had decided to give out a shareholder dividend which was widely criticized. However, this was logical within the premise of a private company. The criticism it provoked, it can be argued, was a sign of the drawing away of support for the shareholder corporate model. But the crisis was not in isolation and had important effects on wider culture of business concerning the appropriate risk-reward but also what UK Plc (UK society) could expect from our companies. This was undermining the broader Thatcherite agenda post-1979 of fostering free markets; increasing managerial authority; increasing the rewards to successful private enterprises; and leaving it up to them how they did that.

Finally, a long-term impact has been the creation of a 'corps' of regulators; who, by regulating a broad section of society have created an inter-related web of regulation; and which, as a result of this web and their time in place, have now developed their own momentum. A community of regulators has quite naturally developed with connections developing between them. For example, in October 1999, they issued a joint statement which included the following: 'Competition provides the best protection for customers and every opportunity has been taken to open avenues to market forces and reinforce competition; Director generals continue to meet regularly. This has created a new powerful body, with its own logic and its own status quo to maintain.'

The role of the regulator

Ofgem, the energy regulator, directly regulates the monopoly elements of the industry – the gas and electricity networks. This legal delineation of politics and regulation masks the complex relationship between the regulator and the political framework since its interface continues to be one of powerful interests, whether business or political. Never forgetting the reality of its complex regulatory and political positioning, Ofgem plays a key role in designing the rules and incentives of the gas and electricity markets, and has a veto on changes to them. Thus, in practice, Ofgem is an extremely important arbiter of the majority of the incentives and rules which affect the value and sale of energy in the UK, and is in that sense central to any energy technology decision, made by any industry actor or customer of any size anywhere in Great Britain. This model of economic regulation, made up of rules, incentives, processes and principles, is central to the transition of the current energy system to a sustainable one. If the design is right it can drive the transition, but if the model is wrong it can act as a major barrier.

This role of Ofgem and its relationship with the key actors relevant to the transition to a sustainable energy economy is discussed in more detail in Chapter 5. However, in brief, Ofgem's role is defined by a set of duties which are set out in the Utilities Act 2000 and the Energy Act 2004. The Utilities Act requires Ofgem to 'protect the interests of consumers, present and future, wherever appropriate by promoting effective competition between persons engaged in ... the generation, transmission, distribution or supply of electricity ...' (Ofgem, 2006b p107). This is its primary duty and requires, amongst other things, that it ensures that licence holders are able to finance the activities which they are obliged to undertake. The Energy Act 2004 adds a 'secondary', and therefore subordinate, duty on Ofgem to carry out its functions in the manner which it considers will contribute to the achievement of sustainable development.

There was not an in-depth discussion at the time of the development of the Utilities Act 2000 of the relationship between Government and the regulator; the extent to which the regulator would be able to satisfactorily manage environmental and social issues; or how they were to be dealt with alongside economic efficiency issues, other than that they should be 'balanced' in some undefined manner. The Labour Government provided Guidance on Social and Environmental Objectives and imposed on Ofgem the obligation 'to have regard to' various socially excluded groups, as well as a requirement to publish a 5-Yearly Social Action Plan and an Environmental Action Plan. Given the absence of a legal ability to impose a hierarchy, Ofgem's primary duty to protect customers through competitive means where possible has continued as the de facto hierarchy. The degree to which any balancing has occurred was where competitiveness, economic analyses or least-cost measures worked with environmental or social goals. Nevertheless, the latter have always been subservient to the economic principle, except where there are legal requirements from Government via Parliament.

The new political paradigm and consequences for sustainability

This book focusses on the degree to which the current political paradigm is suitable for making the move to sustainable development. The central problem has been the broadening of the regulatory oversight to areas which it is not suited to and were never intended for it, i.e. social and environmental issues.

To re-cap on the new political paradigm:

- It is suited to dealing with economic rather than non-economic issues which can derive successful answers from quantitative non-qualitative analyses.
- It is suited to short-term rather than long-term analyses, because of the bullet above.
- It is unable to deal with the complexities of the challenge of climate change, such as the rapid change of understanding of the requirements of climate change or non-economic factors such as human consumption and behaviour.
- It is unable to implement the complex, urgent, linked policies required of a successful innovation process to enable the transition from one energy system to another; e.g. slow progress on renewables, distribution and transmission network development.
- In its efforts to deal with the challenges of climate it necessarily installs policies which can deal with only part of the problem, the economic portion, and which then act as further problems; e.g. the Renewables Obligation, which is having to become more complex in order to overcome its obvious failings but which can never be particularly successful because of its characteristics and therefore is undermining the potential of renewable energy in energy policy.

- It attempts to balance the economic, social, environmental and security goals of sustainable development and energy policy, but necessarily fails because of its economic 'bent', despite the recent inclusion of sustainable development as a criterion.
- And finally, even if it could make these balancing decisions, regulators are inappropriate agents to undertake such balancing, since the decisions are 'political'.

Moreover, there are limits to the degree to which processes of the new regulatory bodies can be effective and efficient. While never intended to be, the 'command and control' nature of network price controls or the implementation of regulated policies are, in effect, what they are: command and control regulations. Command and control has well-known difficulties, which have been argued to be a result of the dilemma of trying to reconcile the links between politics, law, human behaviour and rational economic behaviour (Ayres and Braithwaite, 1992). These difficulties relate to:

- circumvention – when people/businesses try to get around the law, often known as creative compliance;
- perversity – command and control leads to unintended consequences (the level of the Renewables Obligation (i.e. around 70 per cent) which is fulfilled each year maximizes the return to the suppliers, thereby setting up an incentive with the RO not to fulfil the obligation);
- negative feedbacks – which lead to an intensification and complexity of command and control (RO is a classic example);

Furthermore, evolutionary changes – such as market-induced changes with the implementation of the New Electricity Trading Arrangements, and subsequently the British Electricity Trading and Transmission Arrangements (BETTA) – combined with the momentum of the regulatory corps has led to a very particular set of principles, powerfully encapsulated and enunciated by that corps, which (combined with the institutions and political framework in place) focus on certain regulatory outcomes. This focus is strong and powerful and, by their underpinning philosophy or principles, undermines, ignores or excludes large chunks of the energy system, whether it be businesses (i.e. small and medium enterprises, SMEs) or individuals (as citizens or individuals with choice) or institutions (such as local authorities), while at the same time benefitting other chunks of the energy system, namely the incumbents and the status quo.

The 'one size fits all' approach to economic regulation

A key question is whether this 'one size fits all' approach to regulation is appropriate, given the new global concerns of the environment. Joseph, Beesley, Littlechild and all the other early contributors to what became coined as the regulatory state were not writing in a world where climate change was an issue. It was only in 1989, around the same time as privatization of the electricity industry in Britain, that global warming became transformed from a largely scientific issue to a policy problem with the development of global warming potentials (GWPs) (Lashof and Ahuja, 1990). GWPs allowed for the first time an understanding of how important one greenhouse gas was relative to another, and the significance of carbon dioxide.

The principles of RSP are simple and broad enough to be applicable across all sectors, and it is this which both gives the paradigm its strength and the essential characteristic of a new paradigm: its universality. The principles have gradually become codified via legislation within institutions and policies across Britain and have developed into a strong, interwoven web. Both the Conservatives, originally under Thatcher, and then the Labour Government, have reinforced this patchwork. Whatever the sector – telecoms, water or energy, lotteries, the waterways – all are regulated under the same set of principles. Moreover, departments of Government as well as its institutions use the principles of the new political paradigm as boundaries to their policy design and development. The new paradigm, as Keynesian Interventionism before it, has developed a momentum of its own through its institutions and workforce.

The principles of the RSP are applied to widely differing industries, and no doubt all of them have environmental aspects of concern. For example, the communications sector must have concerns about digging up roads and pavements for new technologies, such as broadband, or the effect of millions of old and unwanted mobile phones. However, it is the 'old' natural resource sectors – i.e. water, energy, mining, timber, agriculture – which probably have the gravest environmental concerns to deal with. This is because the issue of sustainability is materially different within these resource sectors, which together form the products in the other non-resource sectors and which form the basis of life – clean water, food, shelter and energy. The goals, and effects, of economic regulation are very different in different industries. For example, innovation and technology development in telecoms or the lottery is very different from the needs, and outcome, of low carbon technology development in the energy sector. Telecoms can be left to competitive forces for technology development. To do this with energy, would not necessarily lead to low carbon technologies, and certainly not quickly enough to meet the challenges of climate change. The thrust of this argument was supported by the Stern Review of 2006 (chapter 2, page 25). It said that climate change is a result of the externality associated with greenhouse emissions – that it entails costs that are not paid for by those who created the emissions. It also argued that climate change has a number of features which distinguish it from other externalities (some of which will be in other sectors):

- It is global in its causes and consequences.
- The impacts of climate change are long term and persistent.
- Uncertainties and risks in the economic impacts are pervasive.
- There is a serious risk of major, irreversible change with non-marginal economic effects.

Energy, therefore, is different. While much of the sector should be regulated for competition and incentives for price reduction as with other sectors, it is a sector which is important in ways that others are not. Because of this, its economic regulation has to be designed to recognize this, and at the very least not make things worse by constraining or channelling technology development.

Principles of the political paradigm

The historical context discussed above has set out what the new regulatory state was trying to get away from and some of the ideas it was trying to institute. From the perspective of Government and the officials within different departments the political paradigm hands down a number of principles:

- Support for competition wherever possible.
- A view that incentives should be designed so that choice is made either by price or by quantity, with the former being preferable but the important point being that policies do not include both of them.
- The cost of policies should, where possible, be minimized.
- Incentives and rules should be technologically and fuel 'blind', meaning that they should not favour a particular technology or fuel (otherwise known as 'not picking winners').

- If an outcome is desired, the design of the policy should where possible mimic competition and should not intervene in the market or alter or undermine incentives elsewhere.
- Analysis should be quantitative and technically expert.

These have a number of fundamental implications which lead to very clear characteristics of the energy system. The implications focus on what types of innovation occur and those that are excluded; the benefits to larger companies, more able to succeed in a competitive arena, and the commensurate dis-benefit to smaller companies; the attitudes to individuals as consumers in a corporate rather than individual sense; and the exclusion of qualitative values and evidence.

Not only are there these implicit implications of the new political paradigm, but also the original intentions have not necessarily been realized. For example, Ministers have not kept out of industry management, as seen from the Railtrack events; or intentions have had unforeseen consequences (such as BETTA and 'queues' for wind power plants to connect to the grid); and finally, new challenges have developed to tax the new paradigm, such as the issues of sustainability. It is not just, therefore, that the design of policies occurs as they do as a result of the constraints placed by the paradigm principles. Increasingly, the challenges the paradigm is expected to deal with are becoming more complex, uncertain and generally more difficult to resolve.

Competition by price

A central principle of the new paradigm is to support competition wherever possible in order to protect the interests of customers. Competition is thought to lead to the 'best' outcomes for society. The principle is that not only do markets make *better* decisions than regulated or interventionist type decisions (for example, decisions of 'picking a winner' by politicians), but by markets making the *right* decisions in the short-term, individual sense, the summing of all those market choices together leads to the 'right' long-term overall outcome.

An integral part of this, is that competition should occur where 'market' rules and incentives are designed to be technology and fuel blind, meaning that there are no rules or incentives specifically intended to 'help' a particular technology or a fuel. This reflects the view that choice by price is the most efficient way to make an individual economic choice and that the sum of those individual economic choices leads to the most effective development of an outcome. Thus, its essence is *not* intend to have a specific technology or fuel outcome, whether channelling outcomes in the direction of sustainability (however that might be viewed) or elsewhere, because that would second-guess the path created from the market choices.

Price or quantity, but don't mix them

A central tenet for policy-making is that prices or quantities can be the means of choice or the desired output to be incentivized, but they should not both occur within a unified policy goal. In other words, they shouldn't be mixed up, potentially confusing incentives or creating 'double-dipping' meaning that some policies receive two incentives rather than one.

Such a view can be analysed in a number of ways, not least from a non-economist's point of view. If something is wanted, why cannot there be more than one incentive and why should they all have the same sort of incentive? Other countries, for example most of those in the EU, allow double-dipping or mixed price and quantity incentives. Why is it that the UK disapproves so much, and why does it matter?

At root, the central point of the political paradigm is that competition should, where possible, occur by price, meaning that the competitive incentive is to reduce price. This is the general catch-all principle of the new paradigm. Where possible, competition is to be supported on price, because it will incentivize the lowest prices, which in turn incentivizes 'innovation' to enable that, which increases economic efficiency. With more money available, changing consumer choice and preferences can then be harnessed to develop new technologies and new innovations based on that further efficiency.

However, in some situations an outcome is wanted to be more 'certain' or 'quicker' than that delivered by a competitive market. An example of this is the desired outcome of a reduction in carbon from the use of energy. Carbon should, in theory, be reduced by a carbon tax. In other words, by setting a price of energy (which includes the carbon tax as a valuation of the carbon externality) and allowing the quantity of carbon emitted derive from that price. However, there are concerns that this will not lead to the 'correct' amount of carbon emitted. At root, Governments are worried that companies will pass the tax on to customers in the form of higher bills and that, because customers have inelastic demand (meaning that they want energy so much that they are prepared to pay whatever they have to in order to have it), the net result will be higher bills without the parallel drop in carbon emissions. This is not a good result for Government, since it would raise the price of energy but without the desired carbon-reducing effect.

Because of concern about this, Governments have so far preferred to place a quantity restriction on carbon, via carbon trading. They prefer this because it means that, in theory, Governments are able to set a 'cap' on emissions, meaning that they know how much carbon will be emitted (or reduced) although they do not know what the cost will be of achieving that cap. Overall though, Governments appear to have decided that if the cost of carbon trading is to be passed on to customers then at least the extra cost will have led to lower carbon emissions. In this sense, they are more certain that they will get the carbon emissions for the higher prices.

Another way to view this policy choice, is that a policy can be established whereby a price is set and this leads to a quantity of something happening. An example of this is the carbon tax as discussed above. Another example, is the feed-in tariff where a price is set and a quantity of renewables is developed as a result of that price. Alternatively, a policy can be set to deliver a quantity of renewable electricity installed or a certain number of tonnes of carbon reduced (for example, the Renewables Obligation or carbon trading) and this leads to a price which is paid to deliver that quantity. That price will rise as the available carbon reduction options become more expensive to harness or as the available sites of renewable energy resources become poorer (and therefore more expensive to harness). The incentive is therefore to find the cheapest sources of carbon or the best renewable resource sites, which is economically efficient.

The UK has policies of both carbon trading and the Renewables Obligation, which is in some ways logical. Most countries in Europe have mixed the policies with both a feed-in mechanism to support renewable electricity but also carbon trading to reduce carbon. It can be argued that the more flexible dual price and quantity policy base is more complementary to the needs of technology development (i.e. building up a new energy system) while at the same time reducing carbon. This shows flexibility in applying appropriate policies for the required outcomes, as argued for by the innovation world (Foxon et al., 2005; Carbon Trust, 2006; Grubb, 2006). This is examined further in a later section in this chapter on the relationship between policies, principles and innovation.

Minimizing cost

Minimizing cost or expenditure by the state by having least-cost policies is economically efficient from a short-term perspective. It is one of those phrases which seems almost 'obvious' – no state wants to spend more than it has to. It is only when the relationship between cost and outcomes is analysed that the statement becomes more complex. Sometimes it might be thought worthwhile spending a little bit more *now*, for something extra *later*. Least-cost has always to be discussed in combination with a timeframe, for example over the next 2–5 years versus the next 50–100 years. A least-cost mechanism in the short term, and the effects of it, will be very different from a least-cost mechanism, and its effects, over the longer term. At root, there is almost no diagreement that least-cost is the preferred option. It is the timing over which any option is valued that is contested; as is what is contained within the calculation; and how that calculation is made.

Short-term least-cost mechanisms complement a 'not-picking-winners' principle. The theory is that a policy or mechanism which is least-cost incentivizes the cheapest outcome or the use of the cheapest technology or fuel. Through the sum of these transactions or choices, the most appropriate outcome (including technology development) will occur. Following a least-cost pathway or a principle of minimizing costs should lead a Government to understand the most appropriate way forward for policy-making. In this way, the combination of least-cost and markets (utilizing competition based on price) will lead to the appropriate 'path' or way forward. In relation to climate change, this principle leads to the view that the reduction of carbon should be incentivized, allowing the market to find the cheapest ways of doing so, rather than the support or 'picking' of specific technologies by Government (discussed in the next section).

Business usually argues for least-cost and broad-based, meaning non-specific, measures so that they are not competitively disadvantaged (domestically or internationally) or so that they do not have to take more responsibility for an outcome than other sectors. An example of a broad-based least-cost mechanism, would be carbon trading. An example of a non-least-cost non-broad-based mechanism might be a feed-in tariff in support of renewable energy. The former example implies that the energy sector does not have to take more responsibility for carbon emissions than, for example, agriculture (since the situation which provides the cheapest carbon reductions will be incentivized, and this could be agriculture, energy or another sector, and it could be undertaking something which direct regulations would not cover). Only markets, it is argued, can incentivize ways of doing things which would otherwise be unthought of. Moreover, other countries are likely to be involved in a broad carbon policy, such as carbon trading. However, since it is an international price of carbon each country has to incorporate similar costs of carbon, and therefore no one country is worse or better off than another.

On the other hand, the outcome of a specific renewable electricity policy is that the electricity industry has to deal with it (and its supporting infrastructure and so on) so that they may have greater costs than other sectors in the economy. Moreover, other competing companies in other countries may have lower costs, so they may be relatively better off.

Parallel mechanisms or arguments in support of minimizing costs in the short term tend to derive from economic cost-benefit analyses. Together, least-cost policies ensure that the cost to UK (or any other country) Plc is as low as it can be so that it minimizes the impacts of competitiveness with other countries. Those industries which do have competitiveness at risk (CAR) issues, are clearly penalized, if domestic policies are not least-cost. Furthermore, if climate change is a global issue (as it is) and if developing countries do not reduce their projected emission pathway (as on the whole they are not expected or required to do for some time) then putting in place specific policies, other than those related to reducing carbon, should, from the point of view of traditional economic theory, penalize ourselves unnecessarily, economically in the short term. The implementation of such policies takes place alongside assessments of 'additionality' or regulatory impact assessments. The basic economic definition of additionality is whether an activity will result in additional benefits over and above those which take place now or are expected to happen anyway (whether, for example, cost, expenditure or additional carbon savings). This value of additionality has provided a way of ranking various initiatives to implement low carbon generating technologies. The central argument of the Stern Review is to undermine these arguments. Nevertheless, it still holds sway over many countries, and remains dominant in the UK.

Technology and fuel blind

A key principle interwoven with being supportive of competition and least-cost policies is to *not* 'pick winners'. In this context, picking winners is taken to mean that market rules or network regulations are *not* technology and fuel 'blind' but incorporate rules which favour a particular technology or fuel. An aversion to 'picking winners' derives from two concerns. The first, discussed above, is that choice by price in competitive markets which are technology and fuel blind is the 'best' way for ordering the outcomes in society and for not competitively disadvantaging businesses. The second concern is that 'directed' choice or 'picking a winner' is a less reliable means of establishing or ordering the long-term preferable outcomes for society or individuals. In practice, not all 'not-picking-winner' policies are 'least cost' – for example, the UK's Renewables Obligation is, in fact, a rather expensive and inefficient mechanism. However, in general, either a least-cost mechanism or a mechanism which mimics market rules is also one which doesn't pick winners.

However, *not* 'picking winners' is sometimes misguided and the concept is in any case misleading. While mistakes of picking an unsuccessful technology are avoided, it can also be that by not picking a winner, a successful technology can be missed. In this sense, it results in avoiding winners. The UK destroyed its nascent wind energy technology base because of the competitive basis of the Non-Fossil Fuel Obligation (Mitchell, 1995). The UK's policy is to reduce carbon dioxide levels by a certain date and to that extent the UK has picked low carbon technologies as the answer to climate change, as it (rather obviously) is. What the current UK Government is not doing, it says, is 'picking' which of those low carbon technologies should be supported, whether it be nuclear power, individual renewable energy technologies, or demand reduction. However, in effect, by not 'picking winners' but deciding that competition by 'price' is the means of choice, the low cost technologies will it is claimed come through. This is, effectively, 'picking' technologies by price. Moreover, that choice is made on a very particular means of establishing 'price' (e.g. usually viewed in the short term) and is also by implication supporting certain technologies more than others: those that achieve rapid price reduction or have less risk attached to the return on investment (UKERC, 2007).

'Not picking winners' and choice on 'price' can be viewed, at best, as an entirely unsophisticated instrument of technological support. It is an 'opt out' from technology and innovation theories, based on empirical studies (Jacobsson and Bergek, 2004). Early and quick price reduction may not mean the technology will prove commercially or even technically viable in the longer term. Ignoring the requirements of technology development while at the same time talking of combating climate change is, at best, the Government putting its head in the sand. At worst, it wilfully ignores the evidence of how different mechanisms of support for innovation work – for example, it takes time for projects to move down their learning curves and this process needs careful nurturing. Cutting them off prematurely when initial results look unattractive (as happened with wave and geothermal energy in the UK in the 1980s) is shortsighted. But you must also expect some losers. That's part of the innovation

process. To knowingly ignore these insights therefore is to knowingly put policies in practice which evidence suggests will be unsuccessful. This is to put policies in place which suit the principles of the political paradigm but which are unlikely to lead to the desired outcomes. In relation to the development of renewables, that decision of non-specific renewable energy policies (which leads to support for the cheaper technologies) versus targeted renewable energy policies (which leads to support of more diverse outcomes) is *the* central choice for the development of renewable energy, according to innovation theory and evidence. The outcome of that decision will lead to very different outcomes in terms of renewable energy delivery, in terms of type of technologies, scale, absolute capacity, investor and customer involvement.

This point can be argued *from evidence* because renewable energy policies have been in place for nearly twenty years and the results of different policies can now be studied. It was understandable, when the first renewable energy mechanisms were put in place in the early 1990s in Europe, that there was limited knowledge of what 'worked' because there were very few mechanisms to 'learn' from. However, this is no longer the case.

> One cannot find a winner without picking some losers: finding solutions to problems requires the research community to explore all reasonable paths in often unknown and risky territories, and inevitably some will be dead ends or 'dry holes'. Thus risks have to be taken; the right strategy is to pull out once an option has been explored and is a proven 'loser'. The [UK] Government has the option of creating a framework of incentives, such as tax credits for RD&D, to devolve the responsibility for picking winners (and inevitably some losers) to industry; but it also has to make choices and take risks too, especially in its support for RD&D where it cannot avoid setting some priorities. The Government has an important role in identifying those of Britain's strengths that are consistent with the industrial environment and the market. It should provide a clear and unambiguous focus. (House of Commons Science and Technology Committee, 2003)

The term itself 'Not picking winners' has become a 'catch-all' phrase used almost as a reason to not 'do' something. Its very universality reflects its bluntness. Of course, there are times when technology and fuel blind market approaches, based on price, are appropriate; and, indeed, they will continue to be the cornerstone of societies' purchasing decisions. However, Governments on the whole put in place policies and mechanisms because

they want outcomes for society which are not occurring when the markets are left to themselves. In other words, because of market failures. The move to a sustainable future requires huge amounts of innovation. The move to the new sustainable economy has to take account of how this innovation can be expected to happen. Economics, competition and price as the key arbiters of choice cannot incorporate these innovation theory findings. As the PIU Energy Review first argued in 2002, and as most European countries accept, such a simplistic principle is no longer appropriate for the ordering of society. 'Not picking winners' is part of an inflexible mindset which is not dealing with the realities of climate change. While it may uphold a political paradigm, it cannot deliver the answers to sustainability.

Implications of the principles of the regulatory state paradigm

There are three major implications of these principles which together leads to the major outcome that there is a constraint on policy design, and this leads to policies which are either ineffective or working too slowly to be of much relevance to meeting the challenges of climate change. The three major implications are: the attitude to innovation; the RSP's natural support for the status quo and large companies; and the RSP's relegation of individuals, citizens, small businesses, local and regional authorities to the sideline of policy design and development.

Attitudes to innovation

The principles of the regulatory paradigm reflect a particular view of innovation and technological development. Chapter 1 introduced the idea of the political paradigm being on the wrong side of the innovation fault-line. The RSP's attitudes are:

- It believes in the linear, controlled development of innovation (which enables a predicted known outcome from policies) rather than a non-linear transition which is based on the idea of the importance of the 'selection environment' to technological development and requires a much broader view of how innovation works including an acceptance that, at best, the direction of innovation can be 'shaped'.
- It considers risk as an important stimulator in innovation while policies which reduce risk, inevitably, soften competitiveness which

in itself must be undermining to incentives which lead to the 'right' answer.

- It considers quantitative economic analyses of markets, innovation and technological development as superior to broad qualitative, as well as quantitative, analyses because it finds the latter difficult to incorporate.
- It considers broad carbon policies superior to focussed, technology policies.
- It supports policies which complement large-scale, status quo companies rather than policies which reach multi-scale, multi-diverse outcomes because according to economic rationale this will lead to the best outcome.

However, with apologies for the brevity and bluntness of the following definition: 'innovation' is taken to encompass the following meaning and requirements. This book takes the view that there is innovation which is enabling the globe to move towards a more sustainable future, and innovation that does not do this. The former is what is 'wanted' and which should be encouraged, to the extent it can be. This 'wanted' Sustainable innovation is not only technological but also political, institutional (including legal, regulatory and planning), managerial, social and cultural. It sums up to be the enabling factors of (or ability to make) the transition from one energy system to another.

It is this mismatch between the needs of innovation and the underpinning principles of the paradigm which is so at odds with the needs of sustainability and is the fundamental reason why the current political paradigm has to undergo radical change. This is covered further in Chapter 3 (which discusses sustainability, innovation and the way forward) and Chapter 5 (which looks at why certain sustainable energy policies have been put in place and why they are so poor at delivering sustainable energy outcomes).

The following sections go through the above bullet points in more detail.

Linear versus selection environment

A central de facto aspect of the current paradigm is the view that innovation occurs in a linear manner. In this comfortable worldview, a policy is put in place and the results can roughly be predicted, occurring in some sort of logical sequence and without unwanted side-effects. This simplifies the policies required to make anything happen, and because outcomes are more or less known a cost-benefit analysis can be undertaken and 'additionality' valued. This, of course, complements traditional economic theory. While market failures are accepted, they can be overcome providing suitable 'shadow' prices or incorporating the value of environmental externalities.

This attitude ignores the idea of a selection environment, which is on the opposite side of the innovation fault-line. The selection environment – conditions in the political, legal, technical, social and economic context of investments – sums up the situation for risk and technological development (Davies, 1996). Investment will only occur if investors believe that the conditions for investment are suitably supportive to whatever it is they want to do. Thus, the selection environment for new entrants, incumbents or different technologies will differ. The current conditions in the energy system's selection environment tend to support larger companies, with access to cheaper financing and economies of scale, rather than smaller companies which have less access. A side-effect of a technology and fuel blind market (which is one factor in the selection environment) is that it is less supportive of new, smaller, diverse entrants, which might (but not necessarily) undertake new and innovative activities. Newer and smaller companies may be more flexible or may be able to survive by finding niches for themselves by doing different activities (such as providing energy services, installing solar water heaters, and so on). These companies will only survive if the selection environment is survivable. If a technology, or a set of technologies, or an outcome, is to be promoted then the selection environment has to be promoted to make it survivable and to build it up into something which can take on a momentum of its own. Larger companies are unlikely to undertake new and innovative activities because in order to stay competitive they are more likely to concentrate on doing the same activities that they already do, but more cheaply. New entrants tend to find niches or areas which the larger companies do not concentrate on. In this way, new ways of doing things develop. A sensible innovation policy attempts to provide some interim market protection for such niche activities, some of which may rely on special local interests, including social rather than commercial interests.

Carbon reduction versus building up a low carbon energy system – i.e. technological versus system

Within the regulatory state paradigm, accepting a linear view of innovation means that a competitive carbon-reducing policy is followed, on the basis that carbon reduction is the goal, and since any one tonne of carbon reduction is the same as another, the most effective policy is one which reduces carbon at the cheapest price. This was discussed above and brings together other paradigm preferences such as minimizing policy costs, least-cost policies, support of competition by price, and so on.

However, in a transition to a sustainable energy system, there needs to be not only a reduction in greenhouse gases (which can be done via least-cost, broad-based policies such as carbon trading) but also the development of a new sustainable energy system. The view of the paradigm is that the development of the 'building up' of a sustainable energy system will also occur as a result of the least-cost broad-based policy. In this worldview, what will happen is that the cheapest means of reducing carbon will be taken up first. Then companies will realize that in order to meet their carbon requirements they will have to move to more expensive means. They will choose which technologies to support based on their best guess of what their best way forward is. The sum of this will lead to the development of appropriate technologies, without picking winners.

This book would argue that this rather stylized view is unacceptable simply from the point of view of urgency. The globe simply cannot wait for the economic system to clunk its way through its processes, bringing forward one technology and then stimulating changes to the legal, planning, electricity standards, infrastructure and all other system changes which are required for each and every technology. Moreover, as this is happening for one technology, so another may be coming forward. It is simply not credible to accept that this process does not constrain or channel technologies but enables them all.

However, even discounting the argument for urgency, this book argues that the 'economic' perspective of technology development is flawed and that a sustainable system needs to be 'built' up. The move to a sustainable energy system requires making the transition from one system to another and requires the basis of the selection environment to move from one system to another. This requires change across and through the selection environment (Rip and Kemp, 1998; Rotmans, 2005; Rotmans et al., 2001; Geels, 2004a, 2005a, 2006). Within economic theory, technologies are enabled as and when they come forward. In practice, some technologies do not come forward and other technologies make it, even if they are not so 'good'. Innovation theory shows how technologies are 'locked in', 'locked out', channelled, and so on (Unruh, 2002; Smith and Stirling, 2006). Nevertheless, the key almost is to keep up with the enormous amount of technological change and to enable its up-take, where this makes sense. It is the urgency of climate change which requires such flexibility and openness to new options.

While carbon-reducing policies may reduce carbon at the cheapest cost, it 'builds' up the sustainable energy system in the matching way of incentivizing the use of the cheapest technologies first, followed by the next cheapest, and so on. In theory, because reducing a tonne of carbon from demand reduction measures should be cheapest, this will occur first, followed by the cheapest, global low carbon supply technologies. To this extent, carbon reductions can be linked through some selection environment change. Since reducing carbon quickly and as cheaply as possible is important, a carbon-reducing mechanism is important. However, broad-based carbon-reduction mechanisms have to be undertaken alongside focussed mechanisms intended to speed up the development, design and operation of the energy system by encouraging change within the selection environment.

However, there are other important issues which should also be considered when looking at the design and implications of any sustainable policy. These include:

- the contribution that the policy makes in system terms (e.g. diversity of supply, contribution to security, location, development of distributed generation options etc.);
- the contribution that the policy makes in terms of overall sustainability (e.g. reduction of emissions, reduced resource depletion, new capacity, the long-term development of new low carbon technologies, possibly issues like size of company, local impacts, contribution to sustainability, including CO_2 reductions);
- the contribution the policy makes to innovation, such as stimulating change or new ways of doing things;
- links with other sectors, such as the renewable energy aspects of agriculture and waste resources; the energy aspects of food policy; and the additional value of 'cross-overs' (or spill-overs) or benefits which occur from one sector to another as a result of a policy (for example, if someone is worried about climate change and is trying to reduce energy, they might also cycle or use public transport, buy local produce, and so on). It is unclear whether one policy leads to this or whether it is the combination of a few, which actually makes change happen.

All these issues are difficult to value from an economic perspective but are without doubt valuable. Conventional economic analyses would value such by an assessment of 'additionality'. The basic economic definition of additionality is whether an activity will result in additional benefits over and above those which take place now or are expected to happen anyway. This value of additionality has provided a way of ranking various initiatives to implement low carbon generating technologies. However, emphasizing the relatively short-term, economic aspects of additionality misses several important areas where some policies make a much clearer contribution to long-term system development and environmental performance. Non-competitive policies may cost more, but they may also lead to the development of new skills (to install or improve that new technology, or, for example, among lawyers or computer modellers); it may stimulate change elsewhere (for example, in network connections); or stimulate additional new technologies to work with that technology (for example, network control technologies). It will have stimulated collateral change, or innovation in unexpected areas.

Risk and innovation – the key issue

Another important aspect of the paradigm is its view of risk and its importance to innovation. On one level, such a view reflects a trust that the cost of this or that technology as stated in a p/kWh assessment by some model somewhere, will be the price at which competition will take place, assuming market failures are dealt with. If markets are set up to ensure choice takes place by price then the most appropriate technology will come through at this price. The innovation world argues that price is just one of several factors (such as institutional, legal, social and political), along with technical compatibility and system integration issues, which are important in the development of a technology (UKERC, 2007; Rotmans et al., 2001; Rotmans, 2005).

For example, wind energy suffers from a number of transaction costs or barriers to investment, thereby making it a greater investment risk than gas plants. These transaction costs include:

- renewables are more capital intensive, so more upfront money per MW is at risk for any particular project;
- higher consent costs per MW;
- many projects don't get the full benefit of economies of scale because they are only 50 MW in size;
- there is a cost associated with the variability of renewables in relation to matching generation to a generator's retail portfolio from hour to hour.

There is clearly a value of 'risk' which is in addition to these transaction costs, which to a degree are measurable. However, it is unclear what the value of this 'risk' is which arises from these transaction costs. The net effect of the combined value of risk and transaction costs is that a generator will only develop, in this case, a renewable energy plant, rather than a gas plant, if they are confident that they will be paid a certain amount more than they would for the gas generation and be confident of a certain return on their investment. This is a very different outcome from what would be understood by comparing the different levelized costs of wind energy and gas generation (UKERC, 2007).

The value of the combined risk and transaction cost will rise or fall, depending on the policy in place because that policy can raise or lower the risk to investment of a project. This is why the choice and design of policies is so important and why, given the current political paradigm, the policies are so poor in delivering sustainable energy policies, because it simply takes no notice of innovation evidence. In essence the fundamental requirements of making the transition to a sustainable energy system are unknown. There are several control and demand technologies within energy – not all of them are fully developed and their best combination is not fully understood. Many of them have different designs and sizes. Some may suit society, or different societies, better than others. The regulatory paradigm says leave it up to the market to decide. And in many ways, this is probably a preferable idea except, unfortunately, there isn't time to allow this to happen. Secondly, anyway, economics picks technologies through price and the lowest price technologies may be those which are 'right' in the short term (e.g. electricity from landfill gas) but not necessarily over the longer term (because of a limited resource) or in conjunction with other technologies. Thirdly, technologies have to be taken up against the momentum of the energy system and the break-out from niches takes time (Raven, 2006; Lamb, 2007). For all these reasons, system transition and technology and innovation literature argues that new technologies need focussed support, adjusted to match the stages in the innovation, early diffusion process and as they develop within markets. This is discussed further in the next chapter, but factoring risk and its effects on investment is an essential requirement for 'opening up' innovation rather than 'closing it down' (Stirling, 2005).

'Buying in' technologies

One view of innovation is the argument that it is economically efficient to 'buy in' a technology once it has been developed somewhere else in the world, that is, paid for by another country. The implication of this is that a new wind turbine or a new wave unit can be 'bought' once it is competitive or near-competitive elsewhere and all the factors which are necessary for it to get up and running (for example, legal requirements, infrastructure, particular standards for energy or electricity production, skills to run it, and so on) will develop alongside it. While the first few projects may be 'demonstration' projects, buying in a technology will enable a country to leapfrog much of the development time and costs. The history of institutional change is ignored in this view. It can be shown that there are real institutional time lags between the beginning of a technology development in a country and an easy 'working' process. If a country waits 20 years before installing a new technology, for example as New Zealand is effectively doing with wind energy, it will still take a considerable time for that 'mature' technology to be matched by a 'mature' selection environment.

Playing to your strengths

It is possible to minimize the risk of 'picking winners' by strategically selecting areas for investment which are best matched to the UK technical and geographical strengths. The House of Commons Committee on Science and Technology (2003) stated: 'The Government seems nervous of being accused of picking winners. As a result tough decisions have been avoided. We should be selecting all of those research projects for funding which we have the capacity to execute and which have a reasonable chance of delivering solutions and significant benefit for UK society.'

Processes of the new paradigm

The process of Government (discussed in the next chapter) or its institutions, for example the process of the regulatory body, can be more or less encouraging to innovation 'in the right direction'. A slow process is likely to be unable to keep up with the complexity and environmental imperatives of climate change and more likely to channel or constrain appropriate technologies, rather than enable them. The purpose of economic regulation within the new paradigm was defined in economic terms. It was important that regulation was not 'command and control', since the market would lead to the appropriate outcomes based on choice and price. However, from the start of the privatizations, and to an increasing degree, economic regulation was legalistic, with a rule-setting and enforcement ethos within an open and transparent, and as a result inflexible, process (Mitchell and Connor, 2002). The combination of this inflexible process, technocratic (read conventional energy system) expertise and economic basis forms a powerful barrier for technology or system development.

The regulatory state paradigm and incumbents

In addition to innovation, the second major implication of the RSP is the way it supports the status quo and the momentum of the current energy system. Competitive markets suit those companies most able to bring together economies of scale and lower costs of capital, since together these lead to lower prices. The principles of the political paradigm can therefore be expected to complement this type of company. As privatization and liberalization have swept Europe, consolidation has occurred across energy companies as they combine to survive in the increasingly tough European energy market. This implies a general presumption that those companies most able to succeed in this competitive environment are beneficial to the UK (or Europe or any other country). If a Government has an energy policy to substantially cut carbon emissions by 2050, as the UK (and European) Government does, and it simultaneously is supporting the status quo by its energy policy, then it also implies that the type of innovation required to move to a sustainable energy system will be incremental and will occur via these large, generally ex-monopoly ex-state companies (or incumbents) in place. It may be that innovation will occur in tandem with these incumbents, but the overall implication is that these incumbents are the important backbone or building blocks of a sustainable energy system rather than a barrier to it. It also implies that the goal of individual country competitiveness takes precedence over other goals, such as the development of SME-type companies.

Competition leads to fewer, stronger, larger companies. In conventional terms, such companies are better able to take on global competitors. Within the electricity industry, the UK's companies are primarily made up of the incumbent ex-monopoly companies, whether they be the electricity generators, energy (gas and/or electricity) suppliers or network companies. There has been consolidation and takeover of the UK ex-monopoly companies by non-UK ex-monopoly companies, such as EDF and E.ON. On one scale, the US is the most productive country in the world and this is because it has the biggest concentration of large companies (OECD Annual Indicators). From the perspective of sustainability, the important question is not whether these large, often incumbent companies are more productive according to an OECD indicator of success, but whether they are able to make the necessary transition to being sustainable energy providers, at the rate required to meet the challenges and environmental imperatives of climate change; and whether they are going to be better or worse than other, possibly new entrant, companies. The whole set of literature on momentum (Hughes, 1983, 1987), autonomy (Winner, 1977), homeostasis (Saviotti,

1986), entrapment (Walker, 2000) and regimes (Rotmans et al., 2001; Geels, 2004a, 2006; Smith et al., 2004) points up the general dynamic wherein incumbents are resistant to change, thus requiring technological innovation to be associated with institutional innovation.

There are further implications of the paradigm principles leading to policies which are complementary to the interests of those companies most able to compete successfully. One is an acceptance by Government of a bias against long-term decisions. Large companies are legally obliged to do their best in terms of shareholder returns. The classic time value of capital will inevitably bias company decisions to the shorter rather than the longer term. A second implication is that Governments are sanguine that the private goals of those companies are acceptably close to the social goals of the UK (Stenzel and Frenzel, 2007).

The corollary of this is that non-large companies, whether SMEs or individuals, are *not* seen as central to combating climate change. This is not to say that the UK Government thinks they are unimportant, but they cannot be seen as central because the fundamental outcome of the paradigm is undermining of them. This an extremely short-term view since SMEs and individuals will be so important in meeting the challenge of climate change.

At root, this is to do with the Government view that it is easier to deal with a few suppliers than it is to deal with millions of customers (*New Statesman*, 2007). And of course, on one level this is another 'obvious' statement. Suppliers have relationships with their millions of customers. It seems 'easier' for Government to put in place policies which enable those few companies to deliver the sustainability agenda. Hence the Renewables Obligation, the Energy Efficiency Commitment, the carbon emission reduction scheme, the proposed supplier obligation – all of which are undertaken via the large ex-monopoly companies.

Domestic customers make the decisions which lead to about 50 per cent of climate change emissions, related to domestic energy use, transport, food policy and waste resources. The large companies, because of their knowledge and their supply base, have the potential to stimulate a great deal of very beneficial innovation and to tap in to these decision makers. The question is whether these large companies will choose to do this; and even if they do, whether their own internal momenta and managerial processes are flexible enough to allow them to stimulate the necessary 'type' of innovation, at the rate which is required. Unless changed, the fundamental incentive on large companies and/or incumbents is to continue doing what gives them the competitive edge – and that tends to be doing the same thing, more efficiently which leads to even lower relative costs. The twin demands of competition and sustainability means that the world of the large companies is complex. They have to succeed in this world. Marketing themselves as 'greener' than the next company, or being 'as' green will be part of their strategy. Thus, change, in general, is acceptable and a normal part of company dynamics and strategy. However, these companies will wish to be in control and able to make the changes they want to make, at the rate they wish. While companies may be prepared to change products or their image, they will not want to introduce products or an image which will undermine their main markets or revenues.

It is recognized what some of the barriers are to moving incumbents towards more sustainable energy practices. For example, it is known that 'market failures' mean that demand reduction measures are not taken up by the domestic sector to the extent that economically rational behaviour would argue that it should be; it is known that network regulation incentives are too orientated (or biased) to increasing capital assets than to performance outcomes (Mitchell and Connor, 2002; Ofgem, 2004b; Woodman, 2007a); it is known that electricity companies are incentivized to sell rather than to provide services (DTI, 2007b); and it is known that network companies are concerned to maintain the ratio of total kilowatt-hours (or therms of gas) sold across their grid (network) to maximum peak capacity (DTI, DETR and Ofgem, 2001; Ofgem 2004b). However, it is not clear that there are reasons why, under the current paradigm, large and/or incumbent energy companies will want to tackle these barriers, given the incentives in place and their other goals. Even if they do decide to implement sustainable energy policies beyond what they would otherwise have done, it is not clear that they will implement them beyond a certain rate.

Moreover, as they become executors of obligations or policies (for example, the Renewables Obligation or the Energy Efficiency Commitment), they become more important to Governments in delivering policies. As a result, they become involved in the development of such policies. It is economically rational that they will be arguing for a design of policy which suits them. They are able to become more powerful, simply because they are so involved. The literature on incumbents and their attitudes to innovation (known as the insider-outsider debate within the innovation literature) was discussed earlier. The conclusion is that incumbents are less likely to innovate, and if they do they will do so at the rate which suits [illegible]ristiansen, 1997, 2003; Christiansen et al., 2004). This has huge [illegible]lly negative and certainly risky) implications for a Government [illegible] arguing on the one hand for a move to a sustainable energy economy and, on the other hand, implements policies which maintain the incumbents' position and strength.

Undervaluing, ignoring or excluding local and individual responses to climate change

The third implication of the political paradigm is the undervaluing of non-national or individual mechanisms to successfully meet the challenges of climate change. It is the other side of the coin to that discussed above: the clear preference Government has on relying on the large companies to deliver the climate change emissions rather than on SMEs or individuals. About half of the UK's emissions come under the Emissions Trading Scheme and, in theory, can be ratchetted down. The other half come from energy use related to transport, domestic households, individuals or smaller businesses, and these areas are also the source of increasing emissions. The current political paradigm effectively ignores the value of individual or local mechanisms to combat climate change. This is more because the greatest thrust of Government principles and policies complements activities of larger companies, thereby strengthening the momentum of the incumbents and the current energy system, thereby making it harder for the transition to a sustainable energy system to take place.

Mechanisms to reduce demand, increase public transport, increase walking and cycling and reduce road transport, increase waste resource use, increase local food sourcing, etc., all revolve around individual or local mechanisms, helped by innovative local authority mechanisms. Their success is largely to do with individual behavioural response. Why individuals do what they do is not well understood, but it is fairly clear that being economically rational and following price signals is only one factor in that understanding. This book argues that while mechanisms such as emission trading schemes are important, mechanisms and attitudes have to be put in place which recognize and target the importance of individuals and small-scale enterprises in tackling climate change. These appropriate policies are more likely to become clear from qualitative rather than quantitative analyses. While large companies clearly have access to these individuals and can undertake innovative and stimulating ideas, it is also important that those individuals are able to take responsibility for their actions in more direct ways, whether buying green electricity, having a micro-wind turbine on their roof which they 'rent' from an energy provider, or being able to buy one themselves from a small, independent company, such as Proven. In conjunction with this, there is evidence to believe that smaller companies or new entrants are likely to be more positive in their stimulation of innovation and changing the energy system, and to this degree measures involving support of the larger companies should always try to 'open up' the system, enabling new entrants (particularly if they then do not have to exist in a competitive process with the incumbents) rather than closing down entry (i.e. Carbon Emission Reduction Target) or maintain it for the status quo (i.e. Renewables Obligation).

The challenge to the regulatory state paradigm

This chapter has discussed the underlying vision of the regulatory paradigm; its principles and their implications. This final section highlights the increasing challenges to it. Sustainability and the increasing understanding of climate change is its focus. This along with the threat to national security are the two greatest challenges facing Governments today. This book does not directly discuss security, except to the extent that a sustainable energy system is able to improve on energy security.

The rise of individualism

As privatization and liberalization have been in place for longer and longer, companies in the licensed and regulated sectors no longer perceive themselves as monolithic. They are increasingly differentiated in terms of size, ownership, geography, technology and strategy. If Government is arguing for a sustainable future, companies will want that to be translated into incentives so that those companies that follow the Government's call will benefit over those that do not. However, the process of economic regulation keeps them herded together because it is to a large degree a system of least common denominators. Some companies may want to be given more of a stake in the future of the industry (in return for responsibilities) and have less to do with the more conventional and inflexible process of economic regulation. However, other companies will not want this. In this situation, the regulator continues either in the middle of this argument or towards the laggard.

Moreover, more companies are becoming involved in the energy sector, with interests in a sustainable energy system, for example via renewable energy, demand reduction, network control technologies. Their numbers are growing at an enormous rate (James and James Annual Reports). They all want more of a 'stake' in the future of the mainstream energy system yet they are rarely able to access that system because of the incumbent generators or suppliers. As a result, these companies remain in the niches on the margins of the energy system. This process of bringing them 'in' could be speeded up.

Similarly, customers also do not want to be viewed as monolithic and increasingly want the ability to fulfil their widely differing choices. Thus, demand and supply customers of all sizes and types are exerting pressure to broaden stakeholder involvement; to change the regulatory process; to take notice of 'future' customers. As customers (of any number, including for example of Tesco) become less and less monolithic; as individual attitudes to sustainability become stronger; and as a 'society' or 'citizen' view of preferred actions becomes more powerful; an ability for these customers to 'choose' is becoming increasingly desirable. These new stakeholders may have different wishes and values than those considered in the UK model of regulation.

In parallel to this, while society is increasing in individualism, there is also an acceptance of, and support for, the importance of public values of trust, fairness and equity, not just nationally but globally. It is less a 'me' sort of individualism as a meritocratic individualism, which also recognizes that for their meritocracy to develop there has to be a wider society based on openness, transparency and fairness. The acceptance of the 'neutrality' and 'value-free' basis of independent regulation is being shaken by that new individualism of society. And this extends to more rigorous questioning of whether the goals of the private companies are the same as, or as good for, society or whether society's goals might differ. To the extent that economic regulation is perceived to be pro-incumbent, pro-big business, 'command and control', anti-choice, pro-monolithic rather than individual, it is increasingly working against the grain of individualism in society.

Increasing complexity of the energy system

There is also a greater understanding of how complex the decisions are facing the regulator, and that they were not recognized in the early days of regulation. In 1990, when the UK electricity industry was privatized, climate change was only just becoming part of the policy language. Now there is questioning whether it is still appropriate that an (economic) regulator should take decisions which will have such an impact on the type of energy system which will develop, given the huge, broad, long-term implications and effects for society of climate change. There is an increased understanding of what a sustainable energy system might be and how different this is in terms of technologies, participants and transactions to the current system. Questions arise about whether the conventional energy system can change to a sustainable system incrementally via its market basis; and even if it can whether it will at the greater rate required to have a positive effect on climate change (Anderson, 2003). The increasing strength and spread of innovation policy-type arguments concerned with momentum, myopia of the 'old' system and transitions force questions of whether economic regulation will constrain and channel a new system, rather than enable it.

As issues become more complex, as more trade-offs have to be made, the regulator has to balance whether the decisions it takes are acceptable within its duties or whether they are outside its scope, and more suited to a political decision. This has the unfortunate effect of further slowing the process of regulation, as Ofgem takes time to think about its position and clarifies its boundaries to the economic sphere, in case it over-steps its duties. Ofgem's natural reaction is to be cautious in case the regulated companies challenge it. Moreover, they will act with their primary stakeholders in mind; and they find it difficult to incorporate values other than economic. The complexities of climate change inevitably are reduced to their economic domain.

This is understandable. However, there are increasing arguments that a narrow concentration on the economic domain is not enough to meet the challenges of climate change. The Stern Review in 2006 argued that while carbon pricing is important, so too are technological development, innovation, consumption and behaviour. It therefore 'matters' if economic regulation, by concentrating on pricing issues and because of its views on innovation, undermines technological development and changing consumption habits.

Transparency and evidence of success

Later chapters will show that the rules and incentives which the paradigm puts in place within markets, networks and policies have a direct effect on the developing characteristics of an energy system. This all reflects an increasing body of evidence of what policies have worked, which have not, why and what their side-effects (benefits and dis-benefits) have been (EC, 2005a; Carbon Trust, 2006; Mitchell et al., 2006; Szarka and Bluhdorn, 2006; Szarka, 2006). And on the whole, these credible studies have shown that the UK policies are inexpensive and inefficient and often have not worked as well as other policies.

There are many different paths or transitions to a sustainable future. Each of these paths implies trade-offs and different benefits and dis-benefits for society. As a result, the policies put in place by the paradigm implies a choice with respect to the trade-offs. A trade-off between the pursuit of economic effectiveness, as measured by a standard economic analysis, and outcomes which tend to arise from non-competitive or non-least-cost policies could be:

- the diversity, and numbers, of new entrants attracted;
- the type of investors (their scale and required returns);
- technology diversity (including scale);
- geographical and resource diversity;
- the development of individual, local and regional benefits, skills and input;
- unknown value added from one sector to another (from energy to waste resource policies) and cross-overs from increased diversity;
- unknown beneficial future innovation.

Some of these outcomes, for example technological, geographical and resource diversity, would lead to lower costs of managing the impacts of intermittent generation (UKERC, 2006). In addition, it may also be that these outputs would lead to more companies, individuals, communities and so on becoming involved in the sustainable energy system, and possibly taking some responsibility for its future. The value, benefits and implications of enabling this are unknowable in an economic way, but certainly require qualitative analyses to have some understanding of their benefits. Denmark, over a 30-year period, has built up a global dominance in wind energy (REN21 Annual Reports). Germany over a 16-year phase now has 240,000 people working in the renewable energy sector (Elliott, 2007a). It is very hard to know what benefits will derive from one, initial small step. The Stern Review attempted this question in relation to the costs of climate change to the UK and came to the conclusion that these indirect benefits are important.

The choice of policy therefore leads to different impacts on society, whether it be for individuals, for small companies, and so on. All 'futures' cannot be enabled and this does in effect require a choice between them. The paradigm has passed this choice on to the market. Even so, choices still have to be made in design of policies according to the constraints of the paradigm principles. It is no longer clear that such choices should be made by the regulator rather than politicians, since the outcomes are so far-reaching in terms of technological futures, society characteristics and implications. The regulatory paradigm has no easy way for these decisions to be made. Thus, the framework of legally separated politicians, and independent regulators which focus on economic objectives which may be able to deal well with certain economic issues, are increasingly unable to adequately deal with complex, fast-moving and long-term problems facing society, such as sustainable development.

Changing relationships

The relationships at the heart of the energy world are also increasingly complex. The relationships between the UK Government and Ofgem; Ofgem and the regulated companies; between Government and wider, bigger energy players, such as BP, are all struggling to find the appropriate balance in a carbon constrained world.

For example, the Government has a complex relationship with the large ex-monopoly companies which were established with privatization and which have developed since then. In principle, the regulatory paradigm would argue that it does not matter who owns the energy industry in Britain or how big the companies get (and therefore how few companies there are, within the realms of an economic appraisal of competition). Instead, the paradigm argues that companies should be judged by their results in terms of efficiency and price.

Moreover, Governments know that companies are required to maximize shareholder interests. Shareholders can call companies to account if they are perceived to be spending money unnecessarily or wrongly. The UK Government appears to continue to 'trust' companies' ability to deliver sustainable outcomes more than individuals or the sum of individuals. However, the Government will also be aware that private interests are by no means allied to social interests with respect to climate change.

The Government is conscious that the major companies are also the main conduits for the low carbon mechanisms, such as the Renewables Obligation or energy efficiency mechanisms, such as the Energy Efficiency Commitment. In these situations, the large companies are effectively in control of the success of those policies, while new entrants find it difficult to be included. Incumbents clearly influence energy policy, given their centrality to these mechanisms. However, this also strengthens the momentum of the current system thereby making it harder to achieve a transition to a sustainable energy system. New entrants, with their often different perspectives, also need to be heard.

A transformation of the relationships between the major actors is required for a transition to occur. In essence though, the current energy system is dominated by a few large energy companies. Their management processes mirror the essential strands of their costs and revenues. If those cost and revenue strands alter, then the management processes of these very large companies have to alter as well. This is no mean feat. Companies will not do this, other than at their own rate, unless they are incorporated into being a 'stakeholder' in the future sustainable energy economy. At the moment, they are the 'child' in the adult–child relationship between regulator and regulated company. The relationships of the energy paradigm are together a part of the 'band of iron', discussed in Chapter 1. They will need to change but it will take changing incentives to do so.

Trust and the government

An increasing challenge for the current paradigm is the falling trust between those that want a sustainable energy system (for example, some academics and NGOs, technologists developing new technologies, new entrants such as sustainable energy generators or demand reduction bodies) because there has been very limited movement forward over the last decade to improve the situation of renewables, combined heat and power (CHP) and demand reduction.

This is partly because of the very 'busy' and changing energy policy of the last decade, culminating in the 2007 White Paper proclaiming the need for nuclear power (discussed in detail in Chapter 4). Despite all this policy debate, very little has actually happened in terms of increased renewable energy installation or reduced demand. For example:

- The interests of 'future' customers, which Ofgem is required to take notice of, are still not explicitly set out seven years after being added to its duties.
- Ofgem has set up initiatives to encourage distribution network operators to innovate (Registered Power Zones and the Innovation Funding Incentive) as well as establishing a premium use of system rate to be paid by distributed generators. However, they are very limited schemes and have not as yet led to increased rates of distributed generation connections (Ofgem, 2003, 2005; Woodman, 2007a).
- Incentives for change in the design and operation of distribution networks are reasonably minor despite a considerable policy debate between 2000 and 2005. The rate of change in the regulation in distribution networks is very slow, so for example, it took five years to adjust engineering standards to ease the connection of distributed generation to networks (Ofgem, 2004b; Woodman, 2007a).
- Access to transmission networks is complex, with some renewable energy generators only being offered contracts for connection in the early 2020s. Moreover, the offshore discussion has been on-going

since 2000 with very little movement forward in terms of rules, incentives and costs of access.
- There are minimal concrete moves towards requiring a shift towards the active management of networks. The lack of movement on active management undermines confidence that Government is serious about distributed generation, and this therefore undermines investment confidence.
- Ofgem is still exclusive, in that it still sees its major stakeholders as those large companies which are network operators or licensed suppliers or generators. It finds it difficult dealing with the wider world of small developers, customers and citizens.

There are growing concerns, as evidenced by the SDC Review of the Role of Ofgem and the DTI's concerns over transmission access (DTI DGSEE, 2007a), that it is not possible to leave the movement to a sustainable energy system to Ofgem's oversight. The 'wish' that this is possible is in large degree because it presents such an enormous task to the Government, both politically and regulatorily, to put in place new institutional measures or legal duties to those currently in place. At some point, it will become clear that not only is Ofgem *not* enabling but *is* constraining the move. Ofgem is simply an institution of the paradigm, albeit now with its own momentum. If Ofgem has to change, so does the paradigm. How the paradigm deals with that conundrum is at the heart of the rest of the chapters of this book.

Conclusions

This chapter has explained what the RSP is; its principles; its attitudes to innovation; and the challenges to it from sustainability. The regulatory paradigm is in many ways alive and kicking, and in many spheres this is entirely appropriate. This book now turns to look in more detail at why the paradigm, in its current form, is no longer appropriate for meeting the needs of the UK, in particular in relation to climate change. Moreover, it is argued that there are increasing tensions at the paradigm's heart, and nowhere is this exemplified more than in its attitude to nuclear power. This is discussed further in Chapter 4. However, the next chapter examines what a sustainable energy system would look like. It also explains how difficult it is to achieve change. Together, these three early chapters explore the concepts, definitions and implications of the regulatory state paradigm. Chapters 4–8 explore what this means in practice both nationally and internationally, and the final chapter puts forward recommendations for change.

3
The Difficulty of Delivering the 'Right' Change Quickly Enough

This chapter explains how difficult it is for Government to create and deliver a new policy. Yet this is clearly much easier than setting in train a series of policies which together would enable the transition from the current to a sustainable energy system. The point of the chapter is to show how a policy change of this magnitude really requires a determined effort by Government. In order to illustrate this, the chapter is made up of three sections. Firstly, it explains what a sustainable energy system is and therefore what kind of change is required within the energy system. Secondly, it explains how Governments 'work' and what the process is for policy development. It illuminates the 'averaging' down or choosing 'the lowest common denominator' which occurs when designing a policy, often because there are just so many competing interests to keep happy. As a result of this, it is difficult to deliver a policy to do something new or different. The role which suits the process of Government is as a deliverer of incremental, slow change. Only when a Government comes in with a landslide is it able to make radical change – hence Keynesian Intervention and the the post-war changes, and Thatcherism and her early policies – or when some shocking event drives new legislation, such as the September 11th attacks in America. Thirdly, the chapter briefly describes the main branches of the innovation literature, and sets out what that literature says are the key requirements of transitional change. It then shows that those policies are not in place. This book makes the argument therefore that a sustainable energy system is very different from the energy system in place today, and that many changes are required to effect a transition. The chapter then shows that delivering new policies which would lead to change is very difficult to do, even for single stand-alone policies within the paradigm. To deliver policies to enable change outside of the paradigm therefore seems even less likely. Finally, the transition literature sets out reasonably clearly what sort of policies or encouragement is required, and these are not occurring in the UK in part because of the paradigm constraint.

The chapter illuminates how hard it is to deliver change, and in particular, the type of change that is needed – in this case innovation which leads to new sustainable energy technologies being taken up, and behavioural and consumption alterations which lead to less energy being used. A sustainable energy system *is* very different from the one in place. It seems unlikely that the necessary changes will occur under the current political paradigm, because:

- the paradigm does not feel comfortable with the types of intervention that are indicated by the innovation literature;
- it does not view innovation as a complex evolving outcome which needs complex and persistent encouragement;
- even if it did see innovation in this way, it is uncomfortable with the types of encouragement that are necessary to move in the right general direction;
- it is unable to adequately understand the qualitative complexities of behavioural change because of its quantitative, economic preferences;
- it is unable to deal with the urgency of climate change.

Together, therefore, the chapter is making the argument that the political paradigm, or landscape as Geels calls it, will have to change if those policies are to be put in place. In other words, the arguments in this book are supported by the transition literature. The implications of this for a paradigm shift are discussed in Chapter 9.

What is a sustainable energy system?

What a 'sustainable energy system' would look like, is widely and hotly debated. The move from the 2003 Energy White Paper (EWP) to the 2007 EWP reflects that. This book would argue that the 2003 EWP reflected one idea of a sustainable energy future which was based on renewable energy, demand reduction, option development and natural gas as a balancer. The 2003 EWP also argued for urgency and, as a result, the need sometimes 'in matters of climate change' to choose the environmental option. This is fundamentally different from the energy system in place. It has taken four years to move from PIU vision and revert back to it.

One aspect of how different actors define a sustainable energy system, is the extent to which they believe 'change' or 'innovation' is necessary to get there. The powerful status quo view is that the energy system does not really have to be very different from that in place today. On the other hand, this book argues that a great deal has to change if the energy system is to be able to make the transition to a sustainable energy future or economy. The former view would argue that there are several paths that a sustainable energy future might take, and this of course is true. The energy system is a very complex and rapidly evolving place technologically, but also socially and behaviourally. There are a great many unknowns which will alter the characteristics of the energy system at different snapshops through time.

However, if the UK were to have similar or slightly higher levels of nuclear power and substantial demand reductions, then the final electricity system would not be that much different from that in place. The configuration of the generators, networks and suppliers may well be very similar. Another example, is that of New Zealand (discussed in Chapter 7) which has 70 per cent of its electricity delivered from hydro power and which has very good renewable energy resources which are only slightly more expensive than the alternatives. It could be argued that the implementation of carbon trading would tip the investment incentives from fossil to renewable energy, thereby enabling a low carbon energy system. Both these examples are technological, non-innovatory and electricity-centric views of energy policy.

A broader view is that even in very unusual systems, such as the Norwegian 100 per cent hydro electricity system, energy system change is going to be required to meet the challenges of climate change outside of the electricity system (for example, with transport) and energy security. Any system with large amounts of direct or indirect fossil fuel use, including for transport or industry, will need to change its characteristics significantly. As discussed below, this requires innovation – 'sustainable' innovation – and it requires it across the various disciplines, institutions and actors, including individuals, that impinge on energy use.

The characteristics of a sustainable energy system

A sustainable energy system is likely to have very different characteristics from those in place in the current conventional energy system. Table 3.1 sets out most of those key differences and shows what is in place today. There are all sorts of unknowns about how the energy system will develop. For example, it is not known which technologies will evolve, or in what way; whether they develop quickly, whether they are taken up by consumers, or whether they fizzle out. It is not known the extent to which obtaining planning permission is going to be a serious long-term constraining factor; whether access to electricity transmission networks will have an important technological blocking effect; whether behavioural issues will affect technological outcomes. There is therefore no one sustainable energy system that has clearly become obvious as the 'right' path, or the path that will be taken, rightly or wrongly. However, it can be expected that a sustainable energy system will embody a number of differences from the current system.

Table 3.1 Differences between the characteristics of conventional and sustainable energy systems

Conventional Electricity System	Sustainable (Low Carbon) Energy System
Attitude to Energy	
✓ Energy there at the flick of the switch	Changing relationship with energy so that there is public awareness of the importance of energy to the environment and the need to use it efficiently
General	
✓ Energy as an input	The need to reduce energy use and to minimize environmental impact act as the basis for innovation
✓ Minimal environmental concerns	Environment is an important driver of policy
✓ Energy security concerns – but answer perceived to be on supply side and to do with conventional technologies	Energy security concerns – but answer to harness diversity of technologies and on both demand and supply side to reduce dependency on gas and by alternative fuels to reduce transport dependency on oil
Minimal social concerns	✓ Social considerations are an important driver of policy
Economic and technology driven	Innovation driven
Technological	
✓ Inflexible electricity generation – coal which needs time to ramp up and nuclear which has to be base load because of its on–off characteristics	Flexible – combined cycle gas turbines (CCGTs) and diverse renewable and distributed heat and/or power technologies
✓ Few technologies and supply dominated	Many technologies – supply, demand, storage and control focused

✓ Reliable – while all plant has generation down-time, in principle able to generate what it says it will	Mixture of output characteristics – reliable, semi-reliable and intermittent
✓ Few, mainly large power plants, connected into transmission system	Many heat and/or power plants of different technology types and sizes, connected into both transmission and distribution network plus self-generators injecting into, or taking from, grid from time to time
✓ Measurement of costs, risk and stimulation of technology innovation neither necessary nor confronted	Measurement of any service, risk, technology innovation or cost fundamental to decision-making
Economic regulation	
✓ Protection of customer interests wherever possible by competition means takes precedence over other Government objectives of energy policy	Although economic regulation still important, its boundary becomes clearer and in matters of climate change, when trade-offs have to be made between Government objectives, the environmental objective takes precedence.
Market and market-rules	
Monopoly and Government owned	✓ Liberalized and privatized
✓ Reasonably simple, undermines innovation	More complex market facilitation between all actors
No choice for *consumers* – few services offered to customers and customers have no obvious means of obtaining what they want	Choice for *customers* multi-services based on both cost and product differentiation offered to customers who are able to make known their wishes
Minimal risk – consumers foot the bill	✓ More risk for companies
When a monopoly, ad hoc regulation by Government and post-1990, RPI-X* mechanism	✓ Increasing proportion of revenue related to performance-based regulation
Costs within network unclear	✓ Clearer costs of using network and providing services, including environmental externalities
✓ Technology innovation and learning curves are unimportant	Technology innovation and learning curves are an important tool in technology choice and business risk
Design and operation of network and system	
✓ Design and operation of network passive and top-down	Operation of network active and multi-dimensional
✓ Based on doing the same thing more efficiently thereby undermining innovation	Incentivizes services and desired outputs, thereby supporting innovation

* RPI-X is the retail price index minus *x* which is a percentage to be set by the regulator.

One clear difference will be the type of technologies which come together to make a sustainable energy system. Currently, most of the sustainable 'supply' technologies are to do with electricity. However, in the future (or ideally as soon as possible) it can be expected that they will expand to include renewable 'heat', for both domestic and industrial customers, and transport. Moreover, there are already many new technologies which sit halfway between supply and demand and are to do with using energy more 'smartly'. This broad range of technologies are often known as 'control' technologies because they enable a different design and use of energy within the energy system; and a different interaction within the electricity system and between systems – i.e. between the gas and electricity system. They also allow integration of different combinations of technologies of different sizes of energy plant and with different characteristics. Thus, the system can expect to have widely different characteristics of scale, diversity, flexibility, resilience, and so on.

The extent to which different technologies develop and therefore which sustainable energy system occurs will have something to do with a number of 'drivers' which exist and are, in a sense, exerting pressure for change and which may, to a lesser or greater extent, influence, enable, channel or constrain the direction of change. For example: the extent, and content, of the research and development programme and the extent to which it enables rather than constrains or channels particular technologies; the economic regulatory system incentives and rules about the operation, design of and access to energy markets and networks, which can enable or constrain and channel the development of different technologies.

Another factor is the extent to which concerns about the environment alter the ability and determination of individuals, customer groups and so on to alter the relationship that they have with their energy use. And then how this may lead on to influence energy companies about the services they offer. Prior to privatization of the UK electricity industry in 1990, customers had no choice and picked up the bill for energy system decisions, and mistakes. Customers have more choice now and the industry increasingly has to answer to their demands. For example, there is now the ability to buy and sell energy services in ways which were not possible in the integrated, pre-privatization world. Nevertheless, customer choice could still be extended greatly, particularly if the rules and incentives in economic regulation change from selling units of energy to providing services as the norm.

Getting from here to there

The language of getting from here to there tends to be that of 'removal of barriers' for new technologies and for new ways of doing things to develop, but it could as easily be about opportunities. This book is arguing that the principles of the political paradigm have to change if these barriers are to be removed and if meeting the challenges of climate change is to be framed in the language of opportunities. It is not that the UK Government does not try to remove the barriers, the problem is that it is so constrained about what policies it can put in place that it is never really able to address the fundamental issues of what has to be done to remove the barriers completely.

An all-encompassing requirement of a significantly reduced carbon energy system is a determination to do things differently from what is in place. This means that incentives which support the momentum of the status quo or the 'carbon' system have to be removed (for example, a preferable renewable energy mechanism would be a feed-in tariff rather than the Renewables Obligation, as discussed in Chapter 5); and new incentives which promote doing 'new, low carbon' things have to be introduced (for example, taking the Energy Efficiency Commitment away from suppliers and opening it up to energy service companies).

The move to new low carbon technologies and new forms of human consumption has to be set in the context of conventional fossil heat and generation which is, on the whole, cheaper than the sustainable electricity and energy alternatives. The conventional technologies, which are on the whole more competitive, are further favoured by electricity market rules because of economies of scale and technology maturity. While ownership of the UK energy systems changed during privatization in 1990, their design and operation continued in much the same way. Market and network rules and incentives reflect competition and the operating characteristics of conventional technologies, favouring incumbents (whether companies or technologies) while the dis-benefits of conventional technologies, for example relating to pollution or inflexibility, are often ignored or not fully internalized. The wider energy system, for example energy for transport, heat for buildings or fuel for industrial processes, also all have their own system momenta. The momenta of the sub-systems are inter-linked and altering all of them so that they sum up to the broad sustainable energy system is complex.

These key barriers, set out below, are the flip-side of the drivers discussed above. Technology development and innovation is never stagnant – it goes on whatever happens elsewhere. However, how it develops is a complex result of the push and pull of drivers, barriers and opportunities, and the way forward will be through the channel which opens up between these different factors.

For example, with respect to renewables, the key barriers are:

- Policy barriers, which encompass many of the others and which this book argues is primarily due to the constraining and channelling principles of the political paradigm in place (Mitchell and Connor, 2002; Mitchell et al., 2006).
- Institutional barriers, which include the new institutions and their principles (put in place by the political paradigm); divided Government aims (e.g. when reducing energy demand is in one Department and supply of energy is in another); divided ministerial responsibilities (e.g. in the DTI when the Minister has responsibilities for different competing interests); those relating to the regulatory environment (governance of the regulator; regulatory duties; relationship between Government and its energy policy objectives and the regulator); and complex but important difficulties, such as those relating to obtaining planning permission for renewables.
- Economic barriers, which reflect a lack of competitiveness for certain renewables related to issues of scale, immaturity of technologies, lack of R&D and demonstration programmes, lack of delivery mechanisms (and constituents of those delivery mechanisms), lack of internalization of external costs, including carbon; lack of appropriate valuing of indirect benefits (such as increased diversity and security of energy system).
- Technological barriers, for example related to the immaturity of technologies or, with respect to renewables, related to the different characteristics which arise from harnessing the wind, the sun, the tides, and so on.
- Social barriers, which arise from patterns of consumption or particular concerns of energy use, such as fuel poverty.
- Infrastructure barriers, which arise as a result of the regulation of the transmission and distribution networks, as discussed in Chapter 5.
- And barriers which arise from energy market-rules and which tend to arise from the principle that market-rules, overseen by the regulator according to its duties, favour no technology or outcome (as discussed in Chapter 5).

As such, the primary building blocks of a 'new' or sustainable energy system would be (and these are developed in Chapter 9):

- Government taking a longer-term view.
- Government viewing climate change as an opportunity.
- The Government being determined to make it straightforward to develop renewable energy.
- Understanding that the transition to sustainability is a 'system' issue.
- When Government is faced with the decision to either encourage diversity and innovation (in other words to 'open up' the energy system) or support the momentum of the current energy system, that it supports the former option.
- An explicit framework for intervention to promote environmental goals, including within economic regulation.
- A focussed, long-term strategy to shift to a sustainable energy system which includes moving from selling units to a service culture; and linking energy with waste resources, transport and agriculture, land use and food policy.
- A low carbon, resource productive economy requires a fundamental change in the attitudes towards energy use and this requires clarifying the roles of the different actors within the energy system, and clarifying the relationships between them.

Expanding on this last bullet, the roles of Government, the regulator, the energy companies and their customers in a carbon constrained world have not been thrashed out. The current relationship between the Government and the regulator, Ofgem, is uneasy. Ofgem is determined to work to its duties and remain independent, but the Government, while supporting this in principle, would also like more 'help' from Ofgem in achieving governmental aims. To a great extent, customers are 'takers' within the energy system; it is very hard for them to become more 'active' and successfully exert choice.

In addition, new relationships between energy service providers, customers, Government and regulators is required. Currently the relationships are (to the extent that so much of the energy world is peopled by men) paternalistic, with the regulator having the final say on what the companies will do. Similarly, Government has the final say in relation to regulations which the companies have to work within. Again, customers are barely involved in any decisions. These relationships have to change in order for energy companies to have more responsibility and be drawn further into discussions for future energy policies (or network development, carbon reduction, energy security). With responsibility should come both a carrot, so that companies gain from moving to a sustainable energy system, but also powerful sticks to deter the misuse of such responsibility.

Customers, whether they be local authorities, businesses, families or individuals, should be able to buy the service they wish; or become involved with energy in ways that they would prefer, for example as self-generators, as investors, as small generators, as purchasers of energy services; or to carry on as they are. Whatever it is that they want they should be able to do; and they should have easy access to information to enable them to do it, and all regulations and standards within the energy system should be changed to enable this.

This boils down to adopting a new attitude towards energy. For example, instead of businesses regarding energy as a factor of production it should be viewed as a means to improve their carbon footprint or to access new opportunities. Instead of Government thinking of it in terms of supply or demand for security of supply purposes, it should be seen as a means to stimulating innovation, thereby enabling society to adapt and move on.

Internal Government process and delivery – the reasons why it is so hard to make anything new happen

The previous section discussed what a sustainable energy system might look like. It described the enormous amount of change which will have to occur if the energy system is to meet the challenges of climate change. This section, on the other hand, examines the internal processes of Government and tries to explain just how hard it is to deliver policy change.

Energy policy has been centre-stage since the Labour Party took power in 1997, though it originally focussed around the re-regulation of the gas and electricity networks via the Utilies Act 2000. A truckers' boycott of oil deliveries in autumn 2000 nearly brought the UK to a standstill. At around the same time, there were several incidences of major electricity network failure in America and New Zealand, raising the fear that this could also happen in the UK. In addition, a Royal Commission on Environmental Pollution report (RCEP, 2000) recommended that the UK set itself on a path to cutting its carbon emissions by 60 per cent by 2050 from 1990 levels.

The Prime Minister asked his own research and strategy group, the Performance and Innovation Unit (PIU), based in the Cabinet Office, to undertake an energy review. The PIU was set up to examine various issues of importance to the Government but outside of the department where the issue was normally based. The typical PIU process was for a team to be put together (made up of PIU officials, officials from other Government parties and external experts) who would work on a short-term project, of about a year. All reports had transparent and clear processes for working, consultation and finalization of the document with relevant departments. The Prime Minister signed them off. The recommendations of the PIU Energy Review represented a real difference in policy, but the subsequent 2003 EWP and 2007 EWP slipped further back to the comfort zone of business-as-usual.

The PIU Energy Review argued that the key to meeting the carbon challenge was to open up low carbon options and that a framework for Government intervention should be agreed upon. It argued that the basis of the latter should be 'that in matters of climate change, when trade-offs need to be made between Government objectives, the environmental objective should take preference' (PIU, 2002 para 3.35). At the time, this phrase slipped through and was not really reported on. In addition to the framework for intervention, PIU argued for increases in the renewable energy targets, large increases for energy efficiency by 2010 and then 2020, and a clearer delivery mechanism for combined heat and power. It also argued that nuclear power was not an appropriate technology to follow at that time. The implementation of such a policy would have represented a fundamental move away from the paradigm principles in place in the UK. The 2003 Energy White Paper was published exactly a year later. It more or less followed the broad vision of the PIU Energy Review, but omitted the framework for intervention. The 2003 EWP was a fudge between setting out a powerful vision for an energy policy suitable for twenty-first-century demands but not actually putting anything in place which would threaten the momentum of the current energy system.

In a sense, a move from the current carbon-based energy system to a low carbon one will only start when the momentum of the current energy system is not only threatened but actively altered by changing the underlying costs, revenues and risks of the incumbent energy companies. Altering those costs, revenues and risks will affect the energy companies' bottom line – their profits – which will have a series of domino effects, throughout the incumbents and eventually to consumers. All the lobbying strength that the incumbents have is to keep these costs, revenues and risks under their control (Stenzel and Frenzel, 2007). They may change them or agree to them being altered, if it is in their interests to do so, but these changes must always occur at the incumbents' pace and in their favour. An analogy used throughout this book is that these underlying costs and revenues are a 'band of iron' holding the current energy system together. Various other aspects of that energy system can alter but it won't fundamentally change unless that band of iron is broken. New ways of doing things will only occur when the incentives change.

Breaking the band of iron of the energy system is a necessary step towards a sustainable energy system, but it is not sufficient to ensure its delivery. This is because the energy system is made up of several sectors and groups, all of which have their own internal momentum. Moving them all forward will be necessary to achieve a sustainable energy system, and for argument's sake, this may include a nuclear future. The energy system itself is made up of conventional energy sources and ways of obtaining those fuel sources and delivering them in usable products to customers. All these segments have their own momentum which strengthens the current system and makes change difficult. From a broader public policy perspective, civil servants who develop, or 'arbitrate' in, policy change are themselves in a system which has its own internal momentum.

When set out in this way, it seems almost impossible that the 'required' changes necessary to enable the challenge of climate change to be met, at society and system level, will occur. Simply achieving *any* change from a public policy point of view is hard, unless pushed by overpowering events such as those of September 11th 2001, and subsequent change to legislation. This section explains why the *process* of Government makes delivering change difficult, even if those changes do not in any way threaten the political paradigm. However, the processes of the political paradigm make it even harder to achieve change if that change does not fit with fundamental principles of the political paradigm. It also explains why as a result of this, policy change tends to occur in incremental steps, building on what is in place before.

Change and the process of government

The difficulty for Governments is that they are having to govern amidst competing wishes of society. While trying to balance all these wishes, they also have to pick up key societal or global issues and 'lead' on them, often in the face of intense lobbying by sectors which don't want change – and incumbents rarely want change. As we all know, stepping back and seeing the 'big picture' of what is really needed in any given situation is difficult at the best of times. And then, of course, Governments themselves want to stay in power. Implementing a policy to deal with a complex issue such as climate change, when there is no universal agreement on what is the right way forward, is always difficult. Doing it within a democratic process with lobbies for and against any given policy, and trying to do it through the processes and aims currently in place, can make it all the more like wading through treacle.

A generalized example of how policy is developed

This section is based on the UK political process, which is based on 'first past the post' elections. Each country has its own political framework, built up over centuries (Clark, 1990). However, most 'developed' countries, whatever their framework, follow a similar process of policy development and delivery, even if the names of the various stages are different. It is the flexibility, or rigidity, of these governmental processes which sets one country apart from another in being able to deliver innovation.

While this section does not attempt a review of the different governmental processes (this is left to Chapters 4–7), it does try to:

- illuminate the 'process' of Government;
- explain how difficult it is for change in policy development and delivery to occur; and
- show that policy development and delivery tend to occur by gradual, incremental change.

The normal procedure of policy development and delivery is first of all to get a 'policy area' accepted in the 'process'. Every democratic Government has only so much 'legislative' time, meaning that there are only so many pieces of legislation which can be be undertaken in any given year or over the term of Government. Because of this, there has to be agreement in what policies are going to go forward to fill that legislative time, and there are formal processes in place to enable this choice. In a sense, there is no point for a Government or govermental department to work on policy development unless it can bring about a desired change. For example, in the UK, the Queen's Speech signals which areas are deemed important enough to warrant legislation and are therefore actively being worked on. However, there will be other areas that are not mentioned in the Queen's Speech but which departments are building up knowledge about, possibly to try for legislation at a later stage. This departmental work falls into various camps:

- policy development approved of and discussed between No.10 (the PM and the No.10/department go-betweens who sit in No.10, work for the PM and are responsible for 'knowing' what each department is up to in detail) and the departments; and

- policy development supported by Ministers or senior officials, but not of immediate importance, particularly in No.10.

Once an area has been highlighted for policy development, the typical procedure is for the drafting of a Green Paper or a policy Review, which is then consulted on. This then leads to a White Paper, which should have taken account of the submissions to the consultation of the Green Paper or the policy Review. The White Paper sets out the policy, but its publication and its policy decisions are often the starting point for change, since each segment has to be negotiated and put into practice.

Green and White Papers follow certain patterns. They are written up by civil servants, and overseen by a Minister. There will be a timetable for publication, which includes a date for a draft to be sent from the 'lead' department to other departments for comments (or several sets of comments) before its final publication. These other departments will then make known any objections they might have to it and any changes they wish to make. There will be a number of drafts prior to publication, each of which takes account of differing departmental views, working towards a final draft which is acceptable to everyone. The lead department is in an important position since it 'holds the pen', meaning it has final say over the wording of the Green or White Paper. The lead department will have to balance the differing comments on the various drafts from the different departments. Many comments are 'cancelled out' by other comments from other departments. Again, most of these concerns are sorted out by civil servants.

However, coming to a final agreement tends to be fairly fraught on a number of levels. Writing a Green or White Paper is a project like any other; staffed by a number of people, who are often very stretched and working to deadlines. If the policy is in any way 'political', meaning if there is interest in the outcome, those individuals will, to an extent, be feeling pressured from competing viewpoints, or if not directly pressured themselves, will understand that some in the team are in that position.

Each department will have worked out its strategy in relation to other departments' wishes. To an extent, the comments on each draft illuminate what each department wants. Making this too obvious may allow other departments to outmanoeuvre them at a later, more crucial stage. For example, Department X might wish to change Line A. Department Y may not want Department X to change Line A. Department X may give up on their demand to change Line A provided Department Y agrees to another change in Line B. The final and most difficult decisions on content will occur at the last moment before publication when the process moves from the 'officials' or civil servant arena into that of the Special Advisers and the Ministers, who have ultimate responsibility for the policy. This is the time when 'to die for', 'non-negotiable', bottom line issues are 'fought' or 'bargained' over by each department. The process itself becomes a mechanism of 'averaging', or rounding off the highs and lows. Fights, skirmishes and boundary-disputes occur between officials and departments over different issues of content; and the Minister – also subject to pressure – has to make a call on the outcome.

However, this is further complicated by the hierarchy of govenmental departments. A policy not supported by the Treasury, for example, is unlikely to survive. On the whole, therefore, very few policies get through which are contentious or particularly different from what has been in place. It is much easier for Governments to do nothing than it is to make change. Even so, it's far easier to do something than to 'kill off' an old policy. And it's far easier to effect change that is complementary to the political paradigm, than against it.

Getting through policies which are contentious requires 'moving to another level' of political negotiation. No political entity is immune from this process, including the Prime Minister. At different times, he or she will have more leverage, and those moments of power will not be wasted but used to push favoured policies. New Governments tend to have been elected on the basis of a few manifesto policies and some of those are more likely to be concentrated on in the first term in office. However, manifesto policies are themselves subject to a political process and may not have support of Ministers. All entities (whether MPs, departments, institutions such as the regulator, NGOs and consumers amongst others) have an amount of power, which waxes and wanes at different times and which always needs to be husbanded.

Government strength is related not only to election votes but also to the election cycle. A Government voted in for a second term on a higher vote is in a powerful position to implement even more radical policies, if they have them, than in the first term. Third term Governments with smaller majorities and wobbly opinion polls face different internal and external policy practicalities. Internal discipline becomes important at this time, and once this discipline has weakened it becomes difficult to agree clear policies, which in turn makes it less likely that the Government will be re-elected. New policies can be developed but the Prime Minister, and other Ministers, will hold on dearly to their 'bargaining chips'. Climate change has had a good outing by the UK Labour Government since

1997, but it has not been the issue on which Tony Blair was prepared to risk his power.

The role of the Minister

The important details of a policy have to be formally agreed by Government. This occurs by the 'lead' department writing a Cabinet Paper, which is sent in the name of the Minister of the lead department to the Cabinet. Efforts are made by officials to work out what Ministers 'want' early on in the policy development process. Officials are able to send the Minister 'briefing notes' to frame the policy debate. Officials, therefore, do have a means of providing information and ideas to Ministers. These types of formal correspondence are what makes up the ministerial 'red box', so often talked about in the media. They also help Ministers to keep tabs on what their (and other) departments are doing. Briefing notes also enable officials to tell a Minister what he or she may not wish to hear but which officials feel the Minister should know. On the other hand, officials do not have to provide information, if the Minister has not asked for it. It may suit officials for the Minister not to know about something. Similarly, a Minister might not ask about certain information, in the knowledge that the answer might not be favourable.

To an outsider, there are few reasons why a Minister has not been told the 'facts'. The reality, when pared down, is close to the *Yes, Minister* TV series. Most of the time, the department trundles along dealing with questions to the Minister; fitting in with the wider political requirements and other department intentions. There are very few key or 'crunch' decisions. It is around these decisions that the *Yes, Minister* situation comes into play. Gerald Kaufman's seminal *How to be a Minister* is still absolutely spot on today.

From a cynical perspective, politicians are on a career ladder like anyone else. They are 'politicians' rather than managers or specialists in this or that subject. The key to their success is the same as for anyone else in any other job. They have to do well at what they are meant to do well at. To the average citizen, we might hope that Ministers who have climate change in their remit would be putting in place good policies for reducing climate greenhouse gas emissions. However, Ministers have their own personal, and career, agendas and will not want to upset someone who might further their office. Moreover, the political party and No.10, as well as other MPs (and factions of MPs), will also have their own agendas. Ministers in charge of the policy have to balance these conflicting demands. Processes of government are put in place to ensure that certain steps are taken, and that, at root, the possibility of corruption is constrained. While checks and balances are introduced to ensure that officials and Ministers act in accountable ways, the processes themselves slow down, and make change more difficult.

The role of officials

Officials (or civil servants) come to their role with their own views and backgrounds, and with varying degrees of spirit. There are individual work concerns, as there are in any work environment: whether they are more successful than someone else; whether they get on with their line manager. An individual will be made responsible for a policy. While developing it, they will be dealing with parliamentary questions, external enquiries, formal meetings, and so on. If the policy is complex or contentious there may be a team which has to be managed, and the individuals in that team will also have their own views and ideas. Thus, policy development is rarely about a 'good' policy but more about working within the current situation – moving whatever has happened before a little bit further forward – and controlling risk to themselves, their department and their Minister. Urgency, passion, interest or commitment to a policy, in a sense, are counter to the Civil Service way of doing things (although, of course, there are many examples of wonderfully committed civil servants).

Officials could be very influential. They write briefing notes, cabinet papers and other advice to Ministers. Ministers are often very busy and they need clear, concise but complete information. Therefore, the degree of effort put in by officials can alter the depth of information provided. Moreover, officials are notetakers at meetings with stakeholders and are the ones who decide which key issues come out of the discussions. Interdepartmental meetings have a process of taking minutes to ensure that participating departments agree on the points raised and decisions taken. Making sure that a meeting is properly minuted can be vital for the development of a policy at a later stage. Provided officials are diligent, this process ensures that the views of different departments are maintained. Not so for stakeholders. Only in formal consultations where submissions are placed on the Web can stakeholders ensure their views are known. However, their interpretation and inclusion in formal documents still depends on the officials.

Taking advantage of, or minimizing disadvantage from, change

Once policy change is under way, a process is kicked off which leads to other changes, some more important than others. From the perspective of Government, and the lead department in the policy process, this is not entirely controllable. The challenge for all interested actors whether

Government, civil servants, lobbyists, NGOs, and so on, is to ensure that their particular area of interest is maintained within the policy development, so that they get want they want out of the process. For the Government, this may, ideally, be a number of outcomes. However, there are probably one or two of the goals which are the most important and if the process starts to become complicated the lesser goals may be dropped. For other actors, this is the time to make sure that their interests are maintained or expanded (or taken out in some situations); that they continue to be closely involved in the process to ensure that this happens; and to ensure that all policy implications are thought through properly. Because of resource and time constraints, the outcome of a policy process is by no means a foregone conclusion.

Renewable energy policy in the UK is a powerful example of this. The first renewable energy policy, the Non-Fossil Fuel Obligation, came into being only because nuclear power was not privatized as anticipated in 1990 and the UK Government had to go to the European Commission to ask for permission to support 'non-fossil fuel', preferring the latter description to 'nuclear power'. One nimble civil servant saw a window of opportunity for renewable energy and took it, arguing that renewable energy was a non-fossil fuel thereby including it in the list of technologies eligible for support (Mitchell, 1995, 2000a). This has had implications for renewable energy policy in the UK ever since. At that time, the UK was clearly in the regulatory state paradigm. The choice was never about a 'perfect' renewable energy policy but simply whether to get any support for renewable energy at all. In the view of this author, the civil servant took the right decision to start the process of renewable energy support in the UK.

Another example, with respect to renewable energy, but this time of being harmed by change is when, a decade later in 2000, the energy industry was partially re-regulated, one of the Labour Party's central manifesto commitments. The policy for renewable energy was a minor issue compared to steering the new Utilities Act 2000 into legislation. Part of the Utilities Act was the section on the New Electricity Trading Arrangements (NETA), which set the rules for how the electricity market worked. Despite the Energy Minister's statements, policies in support of sustainable energy, renewables and combined heat and power were swept aside in the momentum of the development of NETA. NETA (and BETTA and the Utilities Act) are institutions which embody the values of the paradigm. In other words, NETA is a symptom of the problem, rather than the cause of it. In policy terms, the time around 2000 was not so much an open discussion of the pros and cons of different possible renewable energy support mechanisms, but simply one of trying to hold together enough support for *any* mechanism rather than not having one at all (Mitchell and Connor, 2004).

This illuminates just how hard it is for policies that are to be developed in support of something which is not the central piece of any legislation. Renewable energy has never, so far, been a central goal of energy policy. One hopes at some point, UK energy policy does in effect become sustainable, but even with the Climate Change Bill, we are a long way off that.

It also shows how policies which 'fit' the political paradigm are able to be developed and those that oppose the paradigm find it so difficult. John Battle, the Energy Minister at the start of the NETA process, has also been the Opposition Energy Minister. The first 'go' at privatization of the energy industry in 1990 was thought by the Labour Party to be flawed. A central policy of the party's 1997 manifesto was to reform the Gas and Electricity Acts. However, John Battle was soon replaced by various other quick-changing Energy Ministers. It was Helen Liddell who had most influence on NETA's character, although she was neither the Minister in place at its start nor at its implementation. The DTI, in theory responsible for NETA, allowed Ofgem to increasingly take responsibility for the development of its rules and incentives (Helm, 2004; Owen, 2006). With hindsight (and generosity), expecting a Government department to maintain 'intellectual' control over the implementation of something so detailed and so technical as electricity market rules and incentives was probably impossible, given the limited resources it had. Nevertheless, by either losing control, or relinquishing it, has meant that the barriers to renewable electricity and combined heat and power have *increased* under the Labour Government, not reduced.

There are, however, major lessons to be learnt about policy development when large technical projects are involved. The current debate about the value of the NHS is another case in point. These technical projects have major effects on the UK citizen's choices and future. NETA, now the British Electricity Trading and Transmission Arrangements (BETTA), is central to whether we in the UK will achieve sustainable development; but the NHS model impacts all our health lives. As taxpayers, when 'our' Government spends money on one thing, it doesn't spend money on something else.

It is not clear that the DTI had any wish to retain intellectual control over NETA, and was sanguine to allow Ofgem to undertake the day-to-day work of implementation, without doubt clearly to the point of making policy. This is in agreement with the political paradigm principle of 'steering' not 'rowing'. However, Energy Minister John Battle was clear about the outcome he wanted of NETA and it did not go that way. Legally, it was a rather ambiguous time for the relationship between the DTI, Ofgem and NETA because of the development of the Utilities Act. Nevertheless, Ofgem was working to its own current duties and the Minister would not have expected them to necessarily do what he wanted. However, because it was policy development it should have been the DTI which had responsibility. Ofgem should be the executor of policies agreed to by legislation passed in Parliament. The reality is that as NETA progressed, as deadlines became closer and as its rules and incentives were hammered out, it became less and less possible for a supportive Minister to intervene publicly. Finally, sustainable energy policy (renewables and combined heat and power), which initially was thought to be protected within the Utilities Act and NETA process, as stated by Energy Minister John Battle, was sidelined in terms of importance (Owen, 2006).

In the event, Ofgem was in control of the agenda with DTI agreeing to it (Helm, 2004). This policy development had clearly strayed into the realm where Ofgem was making decisions which would have different energy outcomes and which therefore should have been a political decision. Ofgem clearly had an interest that the electricity market rules and incentives were set up as it wanted them to be. However, the final responsibility and 'blame' has to be laid at the feet of the DTI, which should not have lost control. Of course, there are a great many understandable reasons why this happened. The Utilities Act was a very big piece of legislation and no doubt the team developing it was very stretched and under-resourced. Although the renewable energy industry was arguing that NETA could potentially harm it and the Minister was saying he would make sure it did not, it was just too small a side-issue. NETA is a set of rules which reflects the political paradigm. Its rules and incentives are technology and fuel blind, and cost-reflective. Any costs imposed on the electricity system by a generator, or a supplier buying the electricity, are reflected back. This has had major implications for the complexity of calculations of cost-reflectivity for the industry as a whole, and in particular for intermittent generators of electricity, such as wind power. Finally, out of the Utilities Act 2000, came the long awaited renewable energy policy, the Renewables Obligation, in 2002.

The point of this description of the impacts of the development of the Utilities Act 2000 is to explain that unless sustainable energy is at the top of the list of priorities for Government, its policies will be pulled, pushed, squashed and changed to fit other more important priorities. Sustainable energy has not so far been at the top of the list. More or less the same has happened under BETTA, discussed in Chapter 6. Until the environment has precedent over economic concerns, this type of undesirable side-effect can be expected to happen again and again.

System change – domination by convention

This section follows on from the previous two in that it illuminates what the academic energy system transition literature says about the requirements of energy system change (or regime change, as it is called in the literature) and how far this is from the policies currently in place.

Conventional energy systems continue to dominate not only because economics still favour conventional fuel for industrial purposes and generation, but also because of the 'momentum' in the energy system. The term momentum is often used in studies of the development of socio-technical systems, such as the electricity system, to describe the way in which systems develop goals and direction, and to demonstrate a degree of growth suggesting velocity (Hughes, 1983, 1987). Momentum stems from the system's increasingly 'institutionally structured nature, heavy capital investments, supportive legislation, and the commitment of know-how and experience' (Hughes, 1983 p465), and is maintained by the interdependencies between the different technologies used within the system, the manufacturers producing them and the research institutions set up to educate people about the operation of the system, as well as through patterns of Government regulation intended to enable efficient operation. Momentum implies that the system and its characteristics gradually become relatively immune to outside influences, while still possessing the ability to shape the environment in which it operates.

Momentum is often discussed in relation to large technical systems (LTSs). However, other non-technical parts of the energy system also exhibit signs of momentum. For example, the planning system is large and complex with enormous process issues relevant to a transition to a sustainable energy system. Any company or institution could be said to have developed its own internal momentum, and is in large reason why big, powerful companies find it difficult to change to altered circumstances allowing newer, nimbler companies to enter the market successfully. In addition, while the regulatory body, Ofgem, is young, it exhibits characteristics of an internal momentum. This is not surprising. Individuals interview people and choose those which suit the 'principles' of the organization. Even if the 'principles' change, the body will still be employing people who signed up to the original ones. The civil service and large energy companies are in a similar position. The civil service becomes 'attached' to certain policies; attuned to difficulties in policy decision-making; or the likings of a particular Minister or political party.

Individuals also have habits of energy consumption which is then combined with changing technological and cultural choices. The 'kitchen and bathroom' project is a fascinating example of the changing cultural attitudes to kitchens and cleanliness (Shove, 2003). Over a 30-year period, house design has changed considerably to re-focus the house on kitchens and increased privacy and routines of cleanliness. This has had an impact on the amount of energy and water used within the home, as well as resource use to do with kitchen and bathroom 'furniture'. Very little of this is related to price, far more is related to changing behaviour and attitudes to consumption. Quite whether this represents, and should be described as, an increase in momentum of human consumption, an increase in consumerism or a cultural change can be debated. Nevertheless, in general terms at least, human attitudes to energy use are conditioned by the assumption that there will always be enough energy supply to meet energy demand. One aspect of changing the system is in altering the extent to which people take energy's availability for granted. Part of this is connecting people with energy use, so that it does not feel so separate from them and their activities. To the extent that individuals use more energy, changing that momentum of consumption will be of major significance for the success of achieving a sustainable energy economy.

As has been described, the public policy process is very slow moving. Now, add to this the momentum embodied in a large technical system, such as the energy system, which tends to maintain the status quo and makes it more difficult for new technologies to break in. Even with the best will in the world, altering those systems is a major task. Undertaking the kind of change that is necessary to deal with climate change is daunting. Essentially, there are three routes a country can take: it can help these kind of changes along by regulation and altering rules and incentives in favour of a particular outcome; it can leave it up to the market and market mechanisms; or it can have a mix of the two. In practice, most countries follow the third route, but there is a real difference between those that sit at the market end of the spectrum and those that sit at the more regulatory, interventionist end.

This book argues that there has to be a framework for Government intervention to ensure the implementation of policies which deliver a sustainable energy future, by the time it is needed to be in place to make a difference. It is important that 'intervention' is defined and understood for what it is. This book takes it to mean altering the investment incentives for a technology for a period of time by reducing its risk. This might be for a short period of time (a few years) up to a longer, flexible period depending on need. It might mean altering those investment incentives by directly changing rules within the electricity market or network regulation to favour, or exempt, a technology. This might be short to medium term. Intervention can also be understood as being a long-term move to a market where the rules and incentives reflect the characteristics and needs of a sustainable energy system, in the same way that the current paradigm reflects the characteristics and needs of the conventional energy system based on fossil fuels and nuclear power.

Nevertheless, it is important to understand that the successful creation of a sustainable energy system would be one which was an economically efficient market and network system, the output of which was sustainable. The establishment of this energy system and its market and network rules and incentives is taken to be a social construct, just as NETA and BETTA have been. They were constructed to achieve outcomes that are perceived to be important to a (most powerful) section of those involved in its development. However, they could have had very different rules and incentives and the outcomes would have been different. This would not have been any less a 'market', but it would have had different market rules and incentives. Britain, and the RSP, is unusual in the rigour (or inflexibility) with which it excludes intervention. Most countries manage to combine primarily fuel and technology blind markets and networks, with a little intervention in support of sustainable technologies. Such a combination undermines 'pure' economic signals slightly, but only in a limited manner as was discussed in the previous chapter.

Energy system change – what is needed to make the transition

This chapter has tried to explain how much change – on the part of consumers, companies, Governments, institutions, and so on – will be required if the energy system is to become sustainable, and how hard it is to create policies to stimulate or encourage that change.

Altering a large technical system is not going to be easy, although huge changes in technical systems have taken place in the past – for example the shift from horses to the internal combustion engine, or sailing ships to steam ships. The key difference now is that policy has decided that change is vital, and on a short timescale. Earlier changes were, more or less, accidental. If we are to respond to the need to reduce climate change, we have to find some way of bringing a shift about deliberately. This chapter now reviews the various ideas the academic literature has put forward about how change or innovation occurs in society. The idea is to link what the literature says is necessary to enable system change to occur, with the extent to which this is likely to be possible given the underlying principles of the political paradigm. This book has talked about energy systems and political paradigms. But the literature often classifies these as regimes and socio-technical landscapes (Geels, 2004a, 2006). This is confusing on the one hand, but on the other hand reassuring since the transition literature more or less supports what this book is arguing for, even if the terms used are different.

'Innovation' has a huge literature and this review simply picks out the key areas for discussion. A central issue is the importance of customers (or bottom-up type pressures, actions or users) for innovation, as opposed to Government (top-down actions or pressures) approaches. Another key area is the extent to which innovation is considered to be important at all, and if so, whether it is perceived to occur in a linear and predictable manner (discussed below). Another important issue is the extent to which this change or transition can occur incrementally or whether radical or a step-change is required (Ekins, 2000).

This book argues that a whole system approach to system change has to occur and that this will require both top-down and bottom-up approaches. One of the roles of Government therefore is to enable bottom-up approaches to occur. Because innovation is not linear and predictable, it is not possible for Government to put in place policies and 'get' whatever innovation it wanted from that policy. The role of Government is more to encourage innovation by trying to establish environments where innovation may occur. How this can be done is far from clear, and in this sense is contrary to governmental processes which favour economic, technocratic answers.

Linear and predictable

A traditional view of innovation is that a policy or set of policies will set in train certain outcomes, leading to the desired objective. In one sense, all policies have to take the view that their objectives will be met. However, increasingly it is understood that the success of a policy depends to a large degree on the factors which interact with that policy; and that the impact of those factors is inherently uncertain and, as a result, the success of the original policy can never be guaranteed. What this means is *not* that undertaking any policy is hopeless but that when a policy is put in place, or when decisions are made about rules and incentives, that notice should be taken of the somewhat random nature of innovation; the extent to which the selection environment is important; the importance of 'opening up' systems and situations so that the status quo is not maintained and tightened; and the general idea that what is required is an environment conducive to change.

Within individual firms, the conventional view is that innovation activities will take place if there are seen to be sufficient incentives, whether directly financial or driven by the desire to achieve efficiency savings. Innovation activities tend to be path dependent, both because of the established practices and expertise within companies and because the established characteristics guide assessments of what is and is not likely to become a successful innovation. Innovation activities are also guided by the economic rewards available (Kline and Rosenberg, 1986). Moreover, in an established system, incremental innovations are more likely to be deployed because of the interests of established companies and the need to conform to dominant technical standards and system competencies.

Radical innovations which potentially challenge the established characteristics of the system are less likely to be supported by the conditions in the environment in which the system operates – referred to by Nelson and Winter (1977) as the 'selection environment'. They therefore face a range of barriers – institutional, social, technical as well as the 'soft determinism' of economics – and as a result are inherently more risky than a conventional solution.

The economic literature on innovation identifies three main ways in which the performance of an innovation improves over time (Foxon, 2003):

- Learning by doing – firm productivity improves as it gains experience in production which enables improvements in production efficiency.
- Learning by using – the performance of products improves as knowledge is gained from their use in real environments.
- Learning by interacting – product or process innovation comes about by the interactions between producers and users.

However, if an innovation is not being deployed because it is not supported by conditions in the selection environment, then it is not able to exploit these learning opportunities and it will remain an unattractive option for investment. If costs do not fall from learning, the innovation is unlikely to be in a position to achieve the increasing returns of adoption identified by Arthur (1989) as a condition for achieving lock-in.

In an attempt to encourage learning and cost/risk reductions, it is possible to try to protect innovations from hostile conditions in the selection environment. One strategy which may be adopted is to create niches, which can protect innovators from conditions in the system's broader environment and which can therefore act as 'incubation rooms' to allow radical innovations to benefit from learning effects and the creation of supportive actor networks (Hoogma et al., 2002). Niches can emerge fortuitously, as for the gas turbine (Islas, 1999). On the other hand, policy makers can consciously decide to create and manage niches for promising new technologies, a process analysed in the Strategic Niche Management literature (Smith, 2006; Kemp et al., 1998; Kemp et al. 2001; Elzen and Wieczorek, 2005; Jänicke, 2004). The value of strategic niche management is contested to a degree because of its rather linear nature and the extent to which policy makers' actions are assumed to be rational (Berkhout, 2002; Kern and Smith, 2007).

In summary then, the electricity system can be seen as a network of co-dependent technical, institutional, social and political components. A combination of system momentum, economics forces and this co-dependence means that change in the dominant characteristics of the system will be difficult to achieve, and even more difficult at the pace required to reduce the UK's CO_2 emissions. The policies needed to enable the deployment of smaller-scale, lower carbon generating technologies on to distribution and transmission lines will have to provide both sufficient incentives for companies to engage in innovation activities, and to mitigate the conditions in the selection environment to enable the innovation to be deployed successfully. Even so, innovation is an inherently uncertain activity and any new products will need to be able to benefit from learning in order to challenge the lock-in of existing technical solutions.

The valley of technological death

Top-down efforts for innovation can be made from Governments, as much of this book suggests. The argument here is that Governments can do a certain amount but that consumers are a vitally important matching component. Chapters 4–8 look at the ways that Government policies encourage or discourage the move to a sustainable energy system. Governments have a range of means at their disposal to encourage innovation and that transition: they can regulate, provide information, tax and provide financial incentives in the form of support mechanisms, such as the RO or feed-in tariffs but also in terms of capital grants and research funds. This section does not repeat the discussion of support mechanisms – these are explored in Chapter 5 – but concentrates on how focussed research and development (R&D) policies can be very effective in promoting technology development and innovation.

The dispersal of R&D incentives can be made in numerous ways and the Carbon Trust has been very vocal in recent years about the most appropriate way to do this. Targeted support for the early stages of the innovation process can be a relatively low cost and effective way for Governments to steer the innovation process. The Carbon Trust, and others, have broken down the process of technological development from drawing board through to dissemination. The main aim is to avoid any gaps between the different parts of the process. These measures are sometimes described as a combination of technology-push – meaning where measures are pushing the technologies along the innovation process – and market pull, where a market is stimulated for the product which then buys the product and pulls it into the market.

Inevitably some technologies turn out not to be successful. The question is whether technologies should be supported (i.e. picking winners, discussed in Chapter 2) and how long should a technology be supported for? The UK does not have a good history of technology development. For example, the UK did spend money on wind turbine development, but went down the large technology route too early, before the technology was mature enough. It then combined that with an undermining policy mechanism (the non-fossil fuel mechanism) and effectively killed off its industry (Mitchell, 1995). Moreover, the UK closed the wave energy programme in 1992, even though it has now re-opened it. The UK Technology Foresight programme reassessed the wave and tidal programmes in 2001, resuscitating these technologies as options.

Energy system (or regime) transition

The ideas above relate on the whole to specific technologies. This book is interested in how a range of new technologies and ways of doing something develops, alongside all the other factors which are necessary to enable a system transition. Clearly, the system transition is made up of a sum of new products, new disciplines, and so on. A few academic groups or individuals have endeavoured to analyse the requirements of this system-wide transition.

Theorists often argue that the success of a radically different technology requires a change in the underlying technological rationale, which is the set of assumptions about how technology is used and about which technologies are appropriate. These assumptions are a function of the system's momentum and are usually reflected in the social, institutional and economic arrangements and infrastructures that have grown up to support the existing pattern and technological use: the technological regime, or landscape. Technological regime change is seen as necessary to enable (a set of) radically new technologies to flourish by challenging the overall momentum of the system. This essentially is the central argument of this book: energy system change, equals energy regime change, has to occur if a sustainable energy system is to develop.

For example, the current technological regime of the energy system is based on fossil fuels and nuclear power providing large amounts of energy from centralized electricity power plants or reserves of fossil fuels and then transporting it to users. Major corporate and economic interests have emerged to sustain this pattern of operation and support infrastructure, and rules and regulation have developed which support the established technologies, but can act as barriers to new technologies. Raven (2006) argued that there were three dimensions of rules to regimes: regulative, normative and cognitive. Regulative are explicit rules, such as laws and standards and incentive structures. Normative rules include values, norms, and expectations of society. Finally, cognitive rules are those giving meaning to the world, such as priorities, problem agendas, and beliefs. Raven describes how these sets of rules become aligned and thus socio-technological regimes become dynamically stable for a long time, measured in decades.

This understanding of a technological regime helps to explain why changes are typically small, non-radical, incremental and aimed at optimizing rather than transforming the regime. It also explains why so many technologies take such a long time to get from idea to fruition – especially those which require changes in the selection environment, in regulation, in consumer preferences, infrastructure, and so on.

The energy system required for a sustainable energy system is very different, as described earlier in the chapter. Regime change therefore requires the new technologies but also the new support infrastructure, including new institutions. While not every innovation demands a new technological and institutional order, radically different new technologies do tend to be associated with new ways of doing things. For example, the range of social and institutional changes which occurred as a result of diffusion of personal computers and information technology, have effectively decentralized the computer technology. There are parallels to this in the energy system.

Change has to start somewhere and one way to analyse this is to look at how niches begin, grow and develop (Smith, 2006). As discussed above, new products (or companies, ideas, etc.) face difficulty developing because existing products dominate the market, and they are usually owned by a few powerful companies, with vested interests in maintaining the status quo. In this situation, it is not surprising that some new products emerge in marginal and sometimes obscure niche markets, where consumers and users can sometimes play a key role in the innovation and diffusion process. Some niche products, as we have seen, can transcend their initial niches and become mainstream. Examination of these niches has occurred in the hope that if their development can be understood, then possibly some lessons may be learnt about how a sustainable transition can occur. It is not only academics who have analysed this area, but also Governments and companies, to see if they can develop strategically important new technologies. This approach has become known as strategic niche management and the best known Government attempt at incorporating this into policy has occurred in the Netherlands, as discussed in Chapter 8.

The development of niche products has also been thought about as disruptive technologies (Christensen, 1997, 2003; Christensen et al., 2004). Christensen's concept of technological change has some similarities to the idea of regime change. He argues that disruptive technologies require changes in patterns of consumer behaviour and institutional regime change in the wider world. In addition, such technologies are very disruptive for companies, which may have to change their marketing and technology strategy. He argues that disruptive products based on disruptive technologies usually offer lower profits and are most often developed as commercial products in emerging or insignificant markets. A disruptive technology combines lower profits with the least profitable customers in the market (i.e. the innovators or early adopters), and the large mainstream companies have become practised at listening to their best and most profitable customers and hence often miss these disruptive technologies. As a result, Christensen suggests that smaller companies (new entrants) are often in a better position to explore new niches, new technologies, and so on.

As Elliott (2007b) has described, there are examples of community-wide niches developing, for example Woking Council and the island of Samsoe in Sweden. In these large niches, or small energy sytems, the various factors which the new technological regime interacts with have all been changed. The question is how can these large niches or small energy systems be expanded to become a new energy system; or how can enough of them develop so that together they sum to be a new energy system. The reality is that all these factors interact in complex, dynamic ways through the innovation process. Kemp et al. (1998) argue that three strategies are required to effect a regime shift from the niches: a change in economic incentives; that Governments should plan for and build a new regime; and thirdly and most importantly that Governments should 'change the rules of the game'. This means multiple interventions over the short to long term – emphasizing the need for certainty and risk-reduction – as well as the breaking down of the protection of the past technological regime (Lamb, 2007).

Berkhout (2002) argued that the niche to mainstream model of strategic niche management doesn't provide a full explanation of regime change or shift. However, regimes can change and Smith et al. (2004) illustrate technological change in the agricultural sector. They see regime change as being a function of two contradictory processes:

- firstly, the shift of selection pressures acting on the regime;
- and secondly, the co-ordination of resources available inside and outside the regime to adapt to these pressures (which act to keep the regime going).

Policy and actions that seek to address the selection pressures are typically tax or regulatory based. Those that intervene in the innovation system, such as R&D programmes, capital grants and preferential loans, are aimed at shaping regime adaptive capacity. This capacity, or flexibility, is multi-faceted. All other things being equal, regimes with most adaptive capacity will survive in the face of the different selection pressures. It is important therefore if a regime shift is wanted that those two processes work together. In other words, Governments (and other actors to the extent they can) should endeavour to 'open up' the selection pressures (Stirling, 2005) but also, where possible, not shore up the momentum of the current regime.

Socio-technical landscapes

Frank Geels, in a series of publications between 2004 and 2007, opened up an additional level from niches and regimes by adding socio-technological landscapes. He calls the three levels a nested hierarchy. He describes the landscape as a set of deep structural trends, such as economic growth patterns, immigration, predominant political positions, cultural values. Below this are the technological regimes, characterized by incremental improvements, and below them are the niches where radical novelties are generated. This view argues that for niches to lead to regime shift there needs to be a strong alignment of developments at all levels. This complements the arguments of this book that the political paradigm is the equivalent of the landscape; and that the principles of the political paradigm are the equivalent of a band of iron keeping the energy system or regime together.

Niches are where new ways of doing things occur. Regimes select and retain preferred niches. Moreover, regimes are stable because of the strongly interlinked elements, and in stable situations innovation tends to be incremental (Lamb, 2007). If the regime is confronted with changes at the landscape level, the linkages may become looser and actors are able to search for new solutions or new ways of doing things. This creates opportunities for 'niche break-out' (Geels, 2006). New technologies may develop with the old, there are reconfigurations and a chance for more change. In this way, niche applications gradually increase and further reinforce change. Raven (2006) also describes how regime stability may be attacked by changes in the socio-technological landscape with respect to the Dutch electricity regime.

This all sounds terribly sensible and logical. Whereas, of course, in the real world, it is much more complicated. This fits very well with the argument of this book that the political paradigm (or landscape) has to change in order for innovations to be widely deployed and for the system to shift towards sustainability. This offers a problem to policy makers which are trying to stimulate 'good' innovation but it also raises the most difficult question of all, how to achieve a paradigm (or landscape) shift. This is tackled in the final chapter.

Evidence-based analysis of system transformation

Steffan Jacobsson and colleagues have spent the last several years analysing and developing the theoretical basis of technological system transformation. They have worked on the energy system but in particular he and Bergek have analysed how different renewable energy technologies (for example, wind energy and solar systems) have prospered in different countries, such as Sweden, Germany and the Netherlands. Their analysis is based on empirical evidence and has attempted to set out what appear to be the most important factors when comparing the different countries and various outcomes of different renewable energy policies. Their work is to a large degree complementary to the other studies described in this section which have been undertaken on system (or regime) transformation. However, Jacobsson and Bergek differ in that their work is so empirical and they have set out clearly what they see as necessities for successful technological development.

This section now sets out the basic structure that Jacobsson and Bergek (2002, 2004) argue is the necessary foundation for technology development. A technological system is made up of:

- actors and their competencies (may be firms or other organizations);
- networks – as channels for the transfer of tacit and explicit knowledge in both markets and non-markets;
- institutions, which stipulate the norms and rules regulating interactions between actors and the value base of various segments of society.

They differ in this respect from the Institutional Economic literature (Williamson, 2000) which does not highlight the networks. Jacobsson and Bergek then argue that technological systems have five functions, which are not independent of each other so that a change in one may affect another:

- The creation and diffusion of new knowledge.
- The guidance of the direction of search among users and suppliers of a technology – both with respect to growth and legitimacy.
- The supply of resources such as capital and competencies.
- The creation of positive external economies, both market and non-market mediated.
- The formation of markets.

The authors accept that there are two main phases of the evolution of a product or an industry – a formative period and one of market expansion. The formative period is made up of:

- *Market formation* – niche markets (or a series of niche markets which act as a bridge to mass markets) and nursing markets.
- Niche and nursing markets generate space for the *entry of firms* to enter into the value chain.
- *Positive external externalities* occur between firms, such as knowledge or complementary resources and demands, such as legal firms or services, and legitimization.
- *Institutional change* or alignment is at the heart of how new technologies gain ground. Legitimization of a technology and its actors, their access to resources and the formation of markets is strongly related to the institutional framework. If the framework is not aligned with the new technology, several functions may be blocked.
- *Advocacy coalitions* need to be built up so that a range of actors can influence, in competition with other coalitions, institutions.

Niches or nursery markets are the bedrock of this period. New entrants into the market, demanding services and therefore bringing more people in, can together form an advocacy coalition to alter institutions. Thus, niche and nursery markets need to occur early on in the formative stage so that the other stages emerge.

At some point, all of these come together allowing a 'gear shift' in the technological system transformation and begin to develop in a self-sustaining way. This is the stage of market expansion. A necessary condition of this is that larger markets are formed so that the technological system is linked into larger market opportunities. This sets in process a cycle of virtuous circles. However, such a process of virtuous circles will not occur unless the formative period has been completed successfully. Even if it does, the move into market expansion is fraught with difficulty. There are many reasons why making the required investments in this period is risky. Thus, the technological system may develop very slowly. Jacobsson and Bergek give five reasons for this:

- institutions fail to align themselves with the new technology;
- markets may not be formed because (due to increasing returns of adoption) they benefit established technologies;
- (additional) firms may not enter due to lack of markets or because they build on existing knowledge rather than extend their search through new knowledge;
- networks may fail to aid new technology simply because of poor connectivity between actors;
- and because new actors are organizationally weak and unable to counteract institutions or public opinion.

In other words, Governments have to intervene to develop the niche and nursery markets *and* then to enable market expansion. Even so, Jacobsson and Bergek warn that while Governments may establish inducement mechanisms, five major blocking mechanisms become apparent:

- high uncertainty
- lack of legitimacy
- weak connectivity
- ambiguous and/or opposing behaviour
- other Government policy.

In summary, the theoretical and academic literature argues that Government actions are made up of two parallel and overlapping functions. Firstly, the regime and landscape transition literature argues that in order to encourage innovation, Governments have to ensure a conducive environment for it to occur, including the development of niches. The second function is related to the empirical work of Jacobbsson et al. which argues for support for niches and niche and nursery markets, as well as support for technologies during the market formation period. Most strands of the transition literature highlight the non-linear aspect of innovation, although some more than others (Berkhout, 2002; Smith et al., 2004; Kern and Smith, 2007). It is simply not possible to put in place policies which *will* lead to 'innovation'. This book accepts and agrees with this. However, to the extent that innovation is understood, Government should be directive towards Sustainability. What is clear is the UK Government is not doing this (Stirling, 2006b).

Conclusion

Although as we have seen there are a great many theoretical inputs, they do tend to share a consensus in approach, even if the details are different.

1. Consumers are important for successful innovation and diffusion – users can innovate and develop niches which may then move into mainstream; they may also exert influence through their consumer power or through their demands for individualization.
2. Markets and momentum are important since they have such a potentially constraining effect on technologies; but disruptive technologies can require radical changes.
3. Companies and new entrants are important, since incumbents potentially constrain new ways of doing or block niche developments; but new entrants can sometimes break through with disruptive technologies.
4. Governments have to encourage the niches or marginal activities. This means that they have to create a sense of confidence, a sense of long-term support to reduce risk. They should not be shoring up the momentum of the incumbents (as they do); they should try to develop policies which encourage new entrants and new ways of doing things (which they don't); they should be trying to support niches, such as energy services (which they don't).

It is important that Governments provide certainty and persistence of policies, as well as reducing risk. In part, this means endeavouring to create an environment which is both conducive to, and encouraging of, innovation. A full-blooded adoption of regime transitions will not be easy, and perhaps impossible, for the current Government to accept. Regime transition cannot be created to order (as discussed further in Chapters 8 and 9). The momentum of the conventional system is powerful and has resilience to change. If change is wanted then this has to be confronted. Moreover, this is a 'system' issue – in other words, it crosses technological, institutional, social and cultural concerns and is not bracketed solely in one of them. Efforts have to be made to link these areas and focus on a sustainable direction. Engagement with individuals is essential, since a move to sustainable behaviour and consumption cannot successfully be imposed from above.

4
Preferable Intervention – the Pursuit of Nuclear Power

> We appreciate the Government's nervousness about saddling the wrong horse. It would be roundly condemned if it were to put millions into a technology which the market would not support. One need look no further than the nuclear industry for instances where this has occurred.
>
> House of Commons Science and Technology Committee, 2003

Nuclear power has gradually reappeared on the UK's energy policy map. The resurrection of interest in the technology brings it back full circle to where it was in 1979, the last time a Government supported the idea of a programme of nuclear power plants. This resurgent interest comes despite the 2003 Energy White Paper (EWP) which dismissed it on the grounds that its economics made it 'an unattractive option', and that it was an inappropriate generating choice at that time (DTI, 2003).

The foundations for the 2003 statement of Government energy policy were laid in 2000, when the Royal Commission on Environmental Pollution (RCEP) published a report on climate change recommending that the UK should make cuts of 60 per cent in its carbon dioxide emissions by 2050 (RCEP, 2000). In policy terms, this would be an unprecedented commitment from the Government and would go much further than climate change policy in any other country at the time. It is not surprising, therefore, that the Government did not immediately accept the recommendation. However, the RCEP is a statutory body, and the Government was required to produce some sort of formal response to its recommendations.

This need for some sort of response came at the same time as the unexpected but serious, and widely supported, truckers' strike of 2000, over the price of diesel. This, and several major electricity blackouts around the world, raised concerns about energy security and led the Government to form a team of experts within the Cabinet Office to conduct a review of energy policy. During the review but before its publication, the '9/11' assaults on the World Trade Center and the Pentagon took place, further heightening security concerns.

The outcome of the review was a clear recommendation that renewable power and energy efficiency should form the cornerstone of the UK's attempts to cut carbon dioxide emissions and that the Government should accept the RCEP's recommendation for 60 per cent cuts in carbon dioxide emissions (PIU, 2002). The Government used the PIU study as the basis for an Energy White Paper on energy policy which set out its ambition to achieve the 60 per cent cuts by 2050 (DTI, 2003). The 2003 EWP also reaffirmed a domestic political 'target' of reducing carbon dioxide emissions by 20 per cent from 1990 levels by 2010 over and above its Kyoto obligation to reduce its emissions of greenhouse gases by 12.5 per cent from 1990 levels by 2008–12. Like the 1995 Nuclear Review (DTI and the Scottish Office, 1995), the 2003 EWP ruled out public support for new nuclear build, at least in the short term, despite the security of supply and climate change arguments put forward by the industry in their submissions to the PIU. Their decision was based on two factors: the economics of new plant, and the continuing uncertainty about the UK's nuclear waste management programme:

> Nuclear power is currently an important source of carbon-free electricity. However, its current economics make it an unattractive option for new, carbon-free generating capacity and there are also important issues of nuclear waste to be resolved. (DTI, 2003 para 1.24)

Since the 2003 EWP publication, there has been a continuous stream of reports and articles questioning its recommendations and arguing that energy policy, and in particular the issue of nuclear power, should be re-examined (e.g. *New Statesman*, 2007; Rowell, 2006). This argument broadened out to include the Chief Scientist (King, 2005) and the CBI, arguing that the Government should revisit its energy policy (CBI, 2005). The Prime Minister then entered the debate stating that 'the facts have changed over the past couple of years' in relation to energy policy (Blair, 2005). On 29 November 2005, the Prime Minister and the Secretary of State for Trade and Industry announced that there would be a new review of energy to consider all generating options, including nuclear power. Consequently, an Energy Review was published in July 2006 (DTI, 2006a) and a White Paper finally appeared in May 2007 (2007 EWP) making clear its support of nuclear power (DTI, 2007b).

This book argues that this U-turn, which only took four years to come about, enabled the Government to satisfy a number of overlapping, complex policy concerns. Given the twin concerns of security and sustainability, the Government's preference is to support nuclear power which, for reasons explained below, is more in keeping with its political paradigm. In order to increase support for nuclear power it would have to go against some of its fundamental principles, but not many, and certainly less than that required by a decentralized non-nuclear energy policy.

This book argues that the Government knew when establishing its pro-nuclear policy that, at best, it only answers one aspect of the challenges of meeting climate change (a small part of the electricity system) and, at worst, it not only avoids the central issue of climate change (i.e. decarbonizing 100 per cent of the energy system) but that it could make this central issue harder to achieve. In this sense, following the nuclear route is an example of a failure of policy. However, the arguments put forward by the Stern Review in the autumn of 2006 for a secure, domestic price of carbon that was higher than the fluctuating international price, provided a means of supporting nuclear power that also conformed with the political paradigm. Although the Government is likely to have to provide additional support for nuclear power (for example, placing a cap on liable nuclear waste costs for the nuclear industry), it no longer has to intervene in the market place to support nuclear power. Moreover, a domestic price of carbon also benefits renewable energy and demand reduction measures, so it is not 'picking winners'.

Following the nuclear route enables the Government to maintain its view of innovation and remain on its preferred side of the innovation fault-line. Support for nuclear power implies an acceptance of there being a technological, rather than a system, answer to the problem of climate change; that meeting such a challenge does not require fundamental innovation in (or change to) the energy system, since it is more or less a continuation of what is in place; that support for nuclear power complements the view that large companies are the important actors in meeting the energy challenges of climate change; it requires limited responsibility by individuals for their carbon emissions or actions because it puts forward the idea that the answer to climate change is to supply more low carbon electricity rather than to seek to use less energy overall or to consume differently; it implies that the added value which derives from linking sustainable behaviour and consumption across sectors (such as taking public transport, walking rather than driving; worrying about food miles; wearing an extra jumper rather than turning on the heating; having holidays in Britain or Europe, via train) is considered limited. All of which fits with the paradigm.

All in all, the sections below provide a pretty convincing set of reasons to support nuclear power. This book does not accept them and this is discussed in detail later in the chapter. At root, nuclear power is an electricity-only technology which currently only provides 8 per cent of energy supply in the UK. It is the de-carbonization of the other 92 per cent which is important, and this cannot be done by nuclear power but can only be met by a reduction in energy demand; by having a much more efficient energy system; by increasing renewable energy for electricity, heat and transport; and by having flexible technologies to complement them.

There is a wide and deep disagreement about what energy policy should be put in place to meet the challenges of climate change. This is because the two fundamentally different energy policies put forward to answer it (nuclear, renewables and demand reduction versus renewable energy plus demand reduction) reflect two different visions of the future; two different views of what is important in life; two different views of how society should be ordered; and two different sides of the innovation fault-line. In essence, support for nuclear power shows that the current paradigm is still in ascendancy. There are some difficulties for the paradigm in supporting nuclear power if this requires too much intervention and support. However, providing this is not the case, as with a domestic price of carbon, then support of nuclear power fits the paradigm. It shows that the Government is prepared to support a policy which is extremely risky, not just for UK Plc but for the long-term effects on the globe; it shows that the Government prefers to follow such a policy even when it knows this is not the answer to the central problem, just easier to deal with. In this sense, it is a failure of leadership. It also reflects just how deep-set the paradigm is and it does not bode at all well for a move to a sustainable energy system, or indeed any other sort of sustainable system, which requires fundamental system change.

Nuclear power in the UK

The next few sections are heavily indebted to Bridget Woodman's history of nuclear power (Woodman, 2007b). The UK currently has 12 operating nuclear power stations: four Magnox, seven Advanced Gas Cooled

Reactors (AGRs), and one Pressurized Water Reactor (PWR), Sizewell B.[1] The Magnox stations are owned by the Nuclear Decommissioning Authority (NDA)[2] and operated under contract by British Nuclear Group, a subsidiary of British Nuclear Fuels plc (BNFL). The AGRs and Sizewell B are owned and operated by British Energy. Together, these reactors supply around 19 per cent of the UK's electricity, or about 8 per cent of its total energy.

The history of building reactors in the UK is not a happy one. Some AGRs ran significantly over time and budget – most notoriously Dungeness B, where construction took 22 years (Helm, 2004). Even the well established PWR design at Sizewell B ran about a year over its predicted construction time and 40 per cent over budget, with costs rising from £1.8 billion to £3 billion (1993 prices; Thomas, 2002).

As the reactor fleet ages, attention has increasingly become focussed on how existing nuclear capacity will be replaced. Apart from Sizewell B, all of the UK's reactors are currently assumed to have closed by 2023, with Sizewell B scheduled to close in 2035. This closure timetable may well be relaxed if British Energy is successful in its attempts to extend the operating life of its AGR reactors, but in theory it means that around 10 GW of nuclear capacity will be removed from the UK system over the next two decades.

In addition to its nuclear reactors, the UK also has fuel cycle facilities, the most notable of which are the reprocessing and waste management operations at Sellafield. Spent fuel from all reactors except Sizewell B is sent to Sellafield for reprocessing and storage. Low level nuclear waste is disposed of at the nearby Drigg site. Like the Magnox reactors, Sellafield and Drigg are owned by the NDA and operated under contract by BNFL's subsidiary, the British Nuclear Group. Intermediate and high level wastes are stored either at Sellafield or at the civil and military sites where they are produced pending decisions on their long-term future.

The Department of Trade and Industry (DTI) is the lead governmental department responsible for the nuclear industry and energy policy in general. The Secretary of State for Trade and Industry is responsible to Parliament for nuclear safety, the security of electricity supplies and the operation of Ofgem, which regulates the UK's electricity and gas markets. Other departments also play an important role in nuclear matters: the Treasury is responsible for public expenditure and monetary policy, and the Department for Environment, Food and Rural Affairs (Defra) ultimately has responsibility for the industry's environmental impacts as well as climate policy and radioactive waste management issues.

Nuclear safety is regulated by the Nuclear Installations Inspectorate (NII), part of the Health and Safety Executive. Radioactive emissions from nuclear sites are regulated by the Environment Agency, which reports to Defra. The Environment Agency and the NII share responsibility for nuclear waste management.

The history of nuclear power in the UK

In the 1980s, the nuclear industry had high level political support from the Conservative Government under Margaret Thatcher. This came partly from an attraction to the technology itself, and in part from a political belief that a thriving nuclear sector was the best way to undermine the power of the miners in the UK – the Conservatives blamed the 1974 miners' strike for the loss of the General Election that year, and exerted a great deal of political energy in the 1980s in trying to destroy the power of the miners' union (Mackerron, 1996). The Government's support for nuclear generation was cemented by the announcement of a programme of ten new PWR reactors in 1979 (Woodman, 2007b). The case for the first of these stations, Sizewell B, was examined at a public inquiry between 1983 and 1985, and was granted planning permission in 1987. Construction began in 1987 and it was generating in 1994. Another inquiry for a second PWR at Hinkley Point C, began in 1988.

Support for nuclear generation by the Conservative Government was matched by an equal enthusiasm for the separation of public from private; and for an energy policy based on competition between privatized companies (Helm, 2004). But the process of examining the electricity supply industry in preparation for privatization of the electricity industry exposed the high costs of nuclear stations in comparison to coal, in particular because of the reprocessing and waste management costs of the Magnoxes, and also the poor operating performance of the AGR reactors. In addition, the construction of Sizewell B was still in its early stages and any private investor would have to finance a considerable proportion of the ongoing construction costs. These factors meant that nuclear stations would generate electricity at a higher market price than expected. Not surprisingly, then, the Government found that it was unable to interest investors in buying them, and was forced to withdraw the nuclear stations from the privatization programme.[3] In the light of the new information about nuclear costs, it also imposed a temporary moratorium on building further reactors beyond Sizewell B until the conclusion of a review into nuclear power in 1994 (DTI and the Scottish Office, 1995).[4] This meant that the plan set out in 1979 for a fleet of new nuclear power plants, led to just one being built and commissioned by 1994, despite a huge amount of effort spent on trying to deliver more.

Like the 1995 Nuclear Review, the 2003 EWP ruled out public support for new nuclear build, at least in the short term, despite the security of supply and climate change arguments put forward by the industry in their submissions to the PIU.

However, the 2003 EWP did not dismiss new nuclear power completely, stating instead that it would 'keep the option open' (DTI, 2003, para 4.3). In other words, the Government in 2003 was clear that it would not put in place specific measures and subsidies to enable new nuclear build in the UK, although generators were in theory free to bring forward proposals under existing market and regulatory conditions if it wished. No proposals emerged.

Thus, the UK Government has rejected new nuclear power as a viable generating option three times since the privatization of the industry in 1990 (i.e. 1990, 1994, and 2003). There are two inter-related reasons for this rejection: firstly the costs of nuclear generation make it uncompetitive with gas generating options, and secondly the costs and technical issues associated with dealing with nuclear waste make nuclear power politically problematic. These have combined to convince past reviews of the technology that the subsidies required to enable new nuclear stations to be built are not justifiable.

The UK's electricity market is highly liberalized, and without Government subsidy or guarantees, any investors would be required to bear the risks of a nuclear project themselves. These risks fall into several broad categories which extend from pre-construction to final decommissioning and waste management:

- Construction risks: cost overruns or delays in construction have obvious implications for investors and the cost of capital.
- Operational risks: a lower then expected load factor increases the capital cost per unit of output.
- Market risks: the demands of a competitive market and the requirement to operate nuclear stations as inflexible baseload mean that operators are unable to take advantage of fluctuations in the cost of fossil fuel output, and may not be able to achieve a viable price for their output.
- Waste management risks: the high costs of managing nuclear wastes, and the lack of a firm policy in the UK for a final management solution, add to uncertainties inherent in dealing with wastes.

- Regulatory risks: lengthy licensing procedures and the need for proposals to undergo rigorous examination at public inquiries can delay construction timetables.

The potential for any of these risks to be realized in the UK is substantial, given the UK's past construction and operational record with nuclear stations. Combined with this is the fact that any new reactor would be a new design, meaning that there can be little experience of past projects to increase confidence. All of these risks add to the cost of capital, and also to the possibility that the rate of return on any investment in new nuclear build may not be sufficient to compensate for them.

Nuclear waste

As with past reactor construction programmes, the history of radioactive waste management in the UK is troubled and complex. In part this arises from technical issues caused by the reprocessing of spent fuel at Sellafield. The act of reprocessing increases the volume of radioactive wastes which have to be managed, as well as increasing the number of different waste streams. Both these increase the complexity of any final management option.

The legacy of the UK's past nuclear decisions has left the consumer and the taxpayer with an enormous and growing bill. A significant proportion of this is as a result of reprocessing both Magnox and AGR fuel – the most recent Nuclear Decommissioning Authority estimate puts this at around £70 billion (NDA, 2006), although a recent Parliamentary Committee report stated that this is likely to rise significantly (House of Commons Trade and Industry Committee, 2006). In addition, British Energy's remaining liabilities are estimated at around £14 billion (National Audit Office, 2005).

As well as the technical complexities of managing the UK's nuclear wastes, there are also significant political complexities. A large volume of the wastes in the UK will in fact arise from the reprocessing of spent fuel from overseas – a situation which has repeatedly led to the UK being labelled as the 'world's nuclear dustbin' because of a perception that other countries have shipped their own nuclear waste problem to the UK. This has created a particular and possibly unique sensibility to the politics of managing nuclear wastes. Despite this, the generic technical and scientific problems arising from the UK's nuclear waste debate will to a greater or lesser extent become apparent in any other country with a nuclear power programme.

Until the late 1990s, the UK had an established policy of deep disposal for nuclear wastes. Low level wastes (LLW) would either be dumped at Drigg, or stored and disposed of later after the development of an alternative dump site for both low and intermediate level wastes (ILW). High level waste (HLW) is to be stored for 50 years pending the development of a suitable disposal site. This simple description of the policy, however, belies the complex politics of identifying and constructing nuclear waste disposal sites. Several attempts have been made by the industry to push ahead with a disposal site for LLW and ILW, whether in a shallow or a deep facility. All attempts to find a suitable permanent management route have so far ended in politically embarrassing failure.[5] The most recent took place in the mid 1990s when the industry's waste management agency, Nirex, proposed that the first phase of a repository (known as the Rock Characterization Facility – RCF) should be built at its preferred disposal site at Sellafield.

After a planning inquiry, the application to build the RCF was rejected by both the inquiry inspector and the Secretary of State for the Environment. The decision to reject the plan brought the UK's attempts to find a disposal site for LLW and ILW to an abrupt halt after over 20 years of research and nearly £500 million spent on investigations (Parliamentary Office of Science and Technology, 1997).

It would be difficult to overestimate the strategic impact on the nuclear industry of the RCF's rejection. The industry was increasingly presenting itself as a sustainable generating option and as a necessary part of any energy policy seeking to address climate change. The lack of a 'solution' to its nuclear waste problem presents opponents with a powerful argument that an industry which cannot manage its waste should not be considered as sustainable, and therefore that nuclear power should not have a role in the UK's response to climate change. This position was first set out as long ago as 1976 by the politically influential Royal Commission on Environmental Pollution:

> there should be no commitment to a large programme of nuclear fission power until it has been demonstrated beyond reasonable doubt that a method exists to ensure the safe containment of long-lived highly radioactive waste for the indefinite future. (RCEP, 1976)

In addition, with no disposal facility under construction or even in prospect, proponents of new nuclear build find it impossible to give firm estimates about the extent and longevity of the liabilities which will result from the operation of new reactors. This risk will have an inevitable impact on any assessment of the commercial prospects for new build.

The rejection of the RCF left the UK's nuclear waste policy in limbo, where it has remained since 1997. In an effort to build a consensus between the public, politicians and industry, the Government launched a consultation on radioactive waste management with an explicit hope of involving the public in decision-making (Defra, 2001). Its response to the findings of the consultation was to set up an independent body, the Committee on Radioactive Waste Management (CORWM). CORWM's remit has two aims: firstly to propose a technical solution for long-term waste management, and secondly to inspire public confidence in that solution. Over the last two years, CORWM has engaged in a wide ranging consensus-building project to inform its recommendations on future waste management.

CORWM published its final recommendations in July 2006 (CORWM, 2006). The Committee endorses the policy of geological disposal, but, because of the technical and scientific uncertainties about the long-term implications of disposal programmes, it states that it might take several decades to develop an acceptable disposal strategy. In the interim, nuclear wastes should continue to be stored. It must be emphasized, though, that having a strategy or even a firm policy on how to manage nuclear waste is far from having a 'solution' to it. The UK had a clear policy of disposing of low and intermediate level wastes in a deep repository which was to be implemented in the RCF. The policy was effectively overthrown as a result of the decision to reject the RCF in 1997. Whether or not a policy exists, the execution of it can always flounder, and in this case it left the UK without a firm policy on nuclear waste management.

The industry cannot afford another failure in its proposals to deal with nuclear waste if it is to convince investors that new nuclear build is an attractive option. It needs a firm policy in order to be able to estimate its future liabilities, and it also needs assurance that the policy will be put in place and that industry costs will be limited. CORWM may or may not provide a successful step forwards in the development of a feasible waste management strategy for the UK. But in a sense, it is the *perception* of what CORWM achieves which is important in the current debate about nuclear new build. Without a belief that a solution to nuclear waste will ultimately be implemented, it is unlikely that new build will take place.

Nuclear power back with a vengeance

Given this history and the requirements of nuclear power it seems almost unbelievable that nuclear power is back on the political agenda 'with a vengeance' (Blair, 2006a). The 2006 Energy Review announced that 'the Government believes that nuclear has a role to play in the future UK generating mix alongside other low carbon generation options' (DTI, 2006b p124). Any new build will be 'proposed, developed, constructed and operated by the private sector who would also meet decommissioning and their full share of long term waste management costs'. The document commits to enabling pre-licensing of designs and also to resolving the problems faced by nuclear projects in the planning process by severely limiting the scope of the issues that would be considered. These are both issues on which the industry needs to be settled in the short term to allow pre-construction reactor development.

However, the Energy Review stopped short of setting out a detailed plan for achieving new reactor construction in the longer term, although it did provide strong hints about how this might be achieved. Firstly, the Government will work to ensure long-term price stability for carbon within the European Union Emissions Trading Scheme; if this fails, then the UK may take unilateral action to ensure stable prices (DTI, 2006b p34). This would in turn offer a high degree of price certainty for nuclear output. Secondly, the Government and industry will work together to develop a framework for managing long-term nuclear waste costs which appears to be intended to fix the price paid by the industry (DTI, 2006b p123). This would remove much of the uncertainty associated with long-term waste management which has plagued the UK industry since privatization.

While these measures would not be a complete return to overt public sector-led decisions on generation, it would be a significant step along that road. Supporting new nuclear build would backtrack on the increasing liberalization of the electricity sector which has driven the policy agenda since the late 1980s, in that the Government would endorse a single technology (this is not the situation with renewables, where there is an obligation but not a decision on specific technologies).

If it is decided to provide the necessary level of support to enable new nuclear build, nuclear policy in the UK will have come full circle. Privatization exposed the high costs of operating nuclear stations and managing their wastes, and once privatized, British Energy was unable to compete in the market while simultaneously financing its liabilities. New construction has repeatedly been rejected in the liberalized market. Providing support for the industry to build new stations would also mean that the Government will have to accept that its long-standing commitment to the market and liberalization will have to be dismantled.

The factors behind the Government U-turn in 2006/07

The lobby gets going in earnest

The 2003 EWP stated that nuclear power was an unattractive generation choice, but that 'the nuclear option should be kept open'. This phrase, combined with the absence of an expected 20 per cent target for renewable energy by 2020, both incentivized and made it imperative for the nuclear industry to push for new nuclear build. Had there been a 20 per cent electricity target for renewables by 2020 within the 2003 EWP, the situation for nuclear would have been very different. The absence of the hoped-for target was taken to reflect a limited amount of support for renewable energy and undermined confidence in the political commitment to it. A 20 per cent target would have provided certainty of direction of the electricity system, even to those companies not particularly interested in whether the future is renewable or nuclear. It would also have provided confidence to those investors which wanted to support renewable energy but which needed some sign of positive political intent from the Government to reduce their investment risk. It would have reduced the future capacity requirements of the UK electricity system, thereby dampening the incentives for other types of generation and reducing concerns of a 'generation gap'. The price of renewable electricity would have fallen by 2020, thereby making nuclear power unlikely to be competitive with it at that time, and certainly increased the perception of risk in nuclear power. If nuclear power was to have a future in the UK, the 2003 EWP spelt out that it had to get that future *now*; waiting would make it too late.

As a result of this incentive, a rash of articles and papers putting the case for new nuclear stations appeared presenting four basic arguments:

- climate change is such a serious issue that all options to combat it need to be looked at and used together;
- the impending closure of much of the UK's aged nuclear reactor fleet, together with possible future closures of coal plants means that the UK will struggle to meet demand for electricity leading to a 'generation gap', particularly given the urgent need for lower carbon generation and the UK's new status as a gas importer for power generation;
- because of the poor delivery record of renewables and demand reduction, nuclear power is vital to plug the 'generation gap'; and

- the failure of Government policy to deliver the necessary expansion of the renewables industry is evidence that nuclear power is the only option which could guarantee sufficient reductions in carbon dioxide emissions.

While being critical of the performance of renewables and energy efficiency, the majority of these statements assume that the nuclear option complements the continued support and development of other low carbon options (i.e. renewable energy and demand reduction measures), and therefore that all technologies can develop in harmony. The argument that nuclear power is 'complementary' to other low carbon technologies is central to the argument for support for nuclear power. Nuclear power currently provides about 20 per cent of electricity and 8 per cent of energy use in the UK. Since it delivers such a small proportion of total energy in the UK, and is an electricity-only technology, nuclear power can never be 'the' answer to low carbon energy supply. The nuclear power industry knows that all it can be is part of the 'electricity mix', but it has to make the case that it will not undermine the development of other supply technologies or undermine demand reduction which *have* to be the answer.

The Draft 2006 White Paper was accompanied by announcements of Government measures to support renewable energy, no doubt partly in response to the flurry of reports which opposed new nuclear build (Mitchell and Woodman, 2006; SDC, 2007; House of Commons Environmental Audit Committee, 2006). The extent to which these measures actually lead to new delivered renewable energy capacity or demand reduction is discussed in the next chapter. However, whatever the outcome of these new policies it does not undermine the argument that, over time, the financial and political commitment to nuclear power must undermine the development of renewables and the take-up of demand reduction measures. There are several reasons for this:

- Efforts to reduce demand would be less intense so total energy demand, as well as electricity, is unlikely to be contained to the same degree, if the nuclear programme were not supported.
- Since efforts to reduce demand would not be undertaken to the same extent, behavioural issues are not likely to be confronted so that the knock-on effects of increasing waste resources and food policy would also be reduced.
- The supporting institutional framework required to enable new nuclear build would preserve the current economic and regulatory

framework and therefore the current configuration of the electricity industry undermining, and making less likely, the necessary system changes to enable renewable energy and demand reduction.

Competitiveness

The Confederation of British Industry (CBI) was one of the most vocal groups arguing for clarity about energy policy after the publication of the 2003 EWP. The CBI was keen to ensure energy security, meaning enough electricity and natural gas for the needs of its membership, at an acceptable cost. There were concerns that natural gas would become too expensive; concerns of the UK being over-dependent on natural gas for electricity generation, domestic heating and industrial use; that renewables and demand reduction measures were not delivering enough capacity or reduced demand to ensure that a generation gap would not happen; and, finally, that the costs of mitigating climate change would be debilitating to British industry and may undermine British competitiveness as a result of increasing energy prices. The CBI argued for rapid clarification over the place of nuclear power in the energy mix and to get on with supportive measures for nuclear, were that the choice to ensure energy security (CBI, 2005).

This book does not believe that the arguments put forward by the CBI stood up to analysis at the time they were made, and they certainly have not as time has moved on and gas prices have dropped. The Government estimate for the cost of moving to a low carbon economy, which underpinned the PIU Energy Review and the 2003 EWP, was that the cost of moving to a low carbon system with 60 per cent cuts by 2050 was more or less the same as keeping the current business-as-usual energy system going. Moreover, this cost-benefit analysis did not include the benefits from unknown positive innovations, of which there are bound to be some and possibly many. Clearly, in this situation, the obvious Government choice, which it took, was to move to a low carbon economy. Although there have been a number of criticisms of these 2003 EWP figures (Helm, 2004), there have not been any substantial comparable studies producing different (or undermining) figures. On the other hand, there were quite a few studies which supported the 2003 EWP around that time, although for sub-sectors of the economy (WWF, 2005). However, while the costs of not doing anything were argued to be the same over the long term, those costs do not fall equally over time; and they are upfront. Moreover, the 2003 EWP estimates were based on a number of assumptions, and if those assumptions did not work out, then the conclusions would be undermined. To the Government, expenditure on renewables and demand reduction represents a short-term cost to a long-term problem – and this is at the heart of the problem for policy-making.

Since then, there has been the publication of the Stern Review – an elaborate cost-benefit analysis, which takes account of short- and long-term factors, quantitative and qualitative analyses, and a range of disciplines. It argues that the cost to the UK will be far greater if climate change is not confronted immediately. To a large degree, the Stern Review should end the 'competitiveness' debate surrounding climate change. Moreover, the Review also produced powerful arguments in support of a high domestic carbon price, and this is the key to the difference in policies between the 2003 and 2007 White Papers. The Government can set a high domestic carbon price, as called for by Stern, which has the effect of supporting both nuclear power and renewable energy and making it more cost-effective to reduce demand. In this sense, while the Government would have to undertake various other measures of support for nuclear power, it is able to finally get away from direct intervention in the electricity market.

Energy security

Energy security is without doubt one of the two major concerns of energy policy, the other being the environment. Energy security has different qualities, however, since if the lights go off there is an immediate and short-term effect. Climate change is much less obviously visible, and therefore easier to put off from a policy perspective. Security covers an enormously wide area of society, including terrorism or food security. As such, energy security is just one issue within wider security concerns and should be viewed in a system way, in the same way that energy policy is a system rather than a technology issue. In other words, while energy security might be established that would not necessarily mean that UK security is resolved, and indeed it might even deteriorate by concentrating on one particular energy policy.

Energy security is generally thought to cover four areas: that of 'physical' supply, i.e. whether there is enough natural gas available over the short to long term to fulfil the increasing demand world wide; that of markets and whether they will successfully ensure a number of required outcomes, for example ensuring a competitive market for natural gas within Europe or ensuring enough electricity capacity is built in time to remove the 'generation gap'; that of infrastructure and whether incentives are appropriate to make sure that enough investment takes place to ensure that they do not fail; and that of resilience to 'shocks' such as terrorist attacks or major failures (JESS Reports, DTI, 2002 onwards).

A central argument in support of new nuclear build is that it will reduce the need for natural gas in the future and that once supported it will provide capacity making the generation gap less likely. This argument is based on the assumption that once nuclear power is supported, the nuclear power plant will be built. Since the history of nuclear power in the UK presents the exact opposite of this, it seems rather optimistic. But also the argument is made in such a way that the inference is that nuclear power has more chance of being built than renewable energy, and is therefore a more reliable option. In all these ways, nuclear power is therefore argued to be necessary for energy security.

All these arguments put forward in support of nuclear power can be criticized. Certainly, this book does not agree with any of them. Any new nuclear generation would not be built for at least a decade, and possibly a lot longer. Natural gas will still be needed in the short term, and to the extent that nuclear power undermines renewables or demand reduction in the short term, nuclear power plants will make any energy security problem worse because more gas may be required. Support for nuclear power is as likely to undermine investment in other generating power plants, hence raising additional capacity investment problems. With respect to long-term energy security, nuclear power could itself pose an energy security problem. Nuclear power plants are inflexible and, because they are so big and incorporate specific risks, they do not add to resilience in the electricity system. Nor are they flexible. Both resilience and flexibility are recognized as prime characteristics of a more secure energy system.

Finally, the underlying energy policy argument of the 2002 PIU Energy Review and the 2003 Energy White Paper was that natural gas would act as a transition fuel to a low carbon energy system. Its percentage of the market would gradually decrease until it assumes a role as a flexible 'balancer' on the system to complement intermittent electricity from renewables. Thus, natural gas power plants which act as 'balancers' are central to a long-term sustainable energy system. Investing in them combines three necessary functions of a sustainable energy system: low carbon fossil fuel in the short term, flexible capacity additions, and balancing qualities. Nuclear power provides none of these benefits.

Human behaviour and the issues of scale

On the basis of current energy demand, the domestic and transport sectors are proving the most difficult to address in terms of reducing emissions from both a political and policy perspective. If energy demand is not reduced, then the generation gap which needs to be filled would be greater than otherwise thereby making security of supply issues even harder to tackle. The choices made in these sectors are dominated by personal choice. How people consume and why humans behave as they do are also important questions in areas of sustainable development, taken to include sustainable waste strategy, sustainable agriculture and food policy, in addition to sustainable energy. For example, if individuals recycled and re-used more; made choices to ensure that they did not buy excess packaging in the first place; if they bought more locally produced food or grew their own, then energy demand would be very different. The reasons why people consume as they do are very complex.

Other energy policy mechanisms, such as the European Trading Scheme, are aimed at industry and large users. While complex, these mechanisms allow energy use in the industrial sector to be targetted and to be ratchetted down.

From the perspective of Government, reducing domestic and transport emissions is problematic because it requires altering the habits, or affecting the lives in some way, of millions of voters. In effect, Government requires individuals to cut their emissions by 60 per cent if the policy is to cut carbon emissions by 60 per cent by 2050 from 1990 levels. This will both alter, and cut into, the way individuals consume and use energy. This is a major political problem for any Government, particularly if it is already having problems with the electorate for other reasons. This is even more unpalatable to a Government if it thought there was a good chance that the measures it put in place might not be successful. In other words, a Government is only going to do something potentially unpopular if it is sure the plan will work, or if it is absolutely central to its self-definition. Given concerns that voluntary and tax measures may not lead to sufficient emission reduction, there is a strong incentive for the Government to *not* put in place the various measures it has to.

Moreover, increasing renewable energy and reducing energy demand requires myriad renewable energy power plants being built, all of which have to go through planning permission, as well as a roll-out of demand reduction measures to millions of homes. So far, the planning process has been very slow at getting these planning applications through the system.

The issue of scale and numbers of actors involved in the different energy systems is very important within this policy decision. On one level, it is clearly easier if the answer to sorting out the energy policy portion of climate change energy strategy were to successfully build ten or so nuclear power plants. It does not need any fundamental change to the energy infrastructure. The electricity system would still be based on a few energy companies selling units of electricity. It would require either subsidies or an artifically high carbon price, and the latter has been usefully argued for by the Stern Review. It will also require acceptance by Government of waste liabilities over a certain level but it avoids the need to involve the domestic market or individual consumers. It is also an option that many people within the still largely conventional energy industry are familiar with.

In contrast, the alternative energy policy put forward by the 2003 EWP involves developing numerous new technologies which have different characteristics from those of the conventional energy industry. This necessitates new skills, the development of new rules for connection to the grid; upgrades to the transmission and distribution networks; new designs for those networks and new control technologies to enable efficient and secure running of those networks and markets; planning permission for each power plant, when there are difficulties in obtaining them. In order that supply and demand work more closely together, the development of these new supply technologies would link with moves to reduce demand and would increasingly bring in the domestic and transport sectors, involving the many millions of households mentioned above.

While the incumbent electricity companies, including suppliers, could continue to be the major players within this energy system it would be a very different culture for them, requiring for example a move from a sales to a service culture and a move from centralized to a decentralized system. Moreover, this system would enable new entrants at every level, thereby increasing risk to those incumbents. While customers would not have to be involved in their energy decisions, such a system is more complementary to a service culture as it is to micro-generation and it should allow those customers who want choice, or responsibility for their energy footprint, to have it. This is altogether more complex, albeit in many ways far more pro-competition and pro-choice than the current system.

Moreover, if innovation is not thought to be linear and predictable then the role of Government is to try to channel innovation in the 'right' direction as far as possible, and it can do this by encouraging innovation by putting in place policies which stimulate and enable rather than constrain innovation. If the view is that innovation is not important anyway, and even if it were it is linear and predictable, then the Government does not have to act in ways which encourage innovation or try to channel it in the 'right' way. This latter view mirrors the way the Government is acting and, depressing though it is to highlight, this represents a serious lack of leadership.

Slowing down or injecting urgency into the debate?

An additional factor within the debate about the need for nuclear power has been the argument that because climate change is so important, and because it will take so many years to build a nuclear power plant, it is essential that the decision to build new nuclear power plants is taken now. It is often said by supporters of nuclear power that this decision must be taken now; it cannot wait; there is not enough time to wait and see whether non-nuclear low carbon technologies can develop sufficiently over the short term to the extent that nuclear power would not be needed. Because all technologies are assumed to be complementary, there is no down-side to actively supporting nuclear power now. On the contrary, it could even be viewed as an insurance policy (Loughhead, 2005). This argument has somehow managed to pull off what would seem to be the impossible: it has succeeded in making it appear that the Government was not actively doing anything to reduce climate change emissions if it did not support nuclear power, even though there are considerable commitments to expand renewables and energy efficiency; and argue for urgency by supporting a technology which will take at least (being conservative) a decade to get going.

The coming together of the issues and arguments

These hugely contested issues discussed above, and re-capped below, come together to pose a problem for the UK Government:

- that climate change is such a serious issue that all options to combat it need to be looked at and used together;
- that a generation gap is pending;
- that the UK shouldn't become dependent on natural gas;
- that current renewables delivery and demand reduction policies won't offset this generation gap;
- that there is opposition to siting renewables power plants;
- that the central unaddressed issue in dealing with climate change is to do with domestic and behavioural concerns which will be unpopular with voters;
- that climate change is bad for UK competitiveness;
- that radioactive waste can be managed and should not stop nuclear power going ahead; and
- that Government must be seen to be doing something.

The policy outcomes appear to have led the Government to the view that combating climate change effectively at the domestic level and in the transport sector will be very unpopular. Given these difficulties, announcing a return to nuclear power offers a number of tempting benefits. Moreover, Stern has argued for an increase in the price of carbon, so following such a policy would no longer mean 'picking winners' or intervening in the electricity market:

- It will, at best, be over a decade away before a nuclear power plant is built so it puts the problem of combating climate change off while at the same time allows Government to be seen to be doing something.
- Government can state that measures have been put in place to ensure the generation gap will not appear and therefore energy security fears can be allayed.
- There will be more time for the other alternatives, renewables and demand reduction, to expand.
- There will not be such a pressing need to urge demand reduction, so politically unpopular policies won't have to be put in place.
- Pressure for successful planning applications for renewables will be reduced.

Support for nuclear power – principles to put on hold

The sections above provide a pretty convincing set of reasons, albeit highly contested, to support nuclear power. But the problems are that:

- Nuclear power is not complementary to renewable energy and demand reduction (which have to be the answer for a low carbon energy system) and will therefore undermine their development.
- Once a policy to support nuclear power is embarked upon, nuclear power plants have to be built otherwise the UK will not be able to meet its domestic or international commitments to carbon reductions. Given the history of nuclear build in the UK and more recently Finland (TVO, 2006), this is a very risky strategy.
- Nuclear power is a small part of the problem – only 8 per cent of energy supply in the UK – so that concentrating so much time and effort on its development is not concentrating on the central issues.
- The Government could have more renewable energy capacity and demand reduction quickly if it implemented different policies.

- No effort is being made to address the challenge of climate change as a system problem which the 2003 EWP tried to do and which the 2007 EWP turned its back on.

Moreover, in order to support nuclear power the Government has had to go against some of its fundamental principles, but not many, and certainly fewer than that required by a decentralized non-nuclear energy policy.

The promotion of competition

'Choosing' nuclear power ignores a central tenet of the regulatory state paradigm: to not pick winners. Malcolm Wicks was the Energy Minister from early 2004 to late 2006. He repeated several times that nuclear power would have to stand on its own two feet and that Government would not subsidize it (Wicks, 2006). This has been met with, at best, scepticism.

The clear message which has come across about nuclear power since the Conservative Government tried to privatize it in 1990, is that it is not a technology which suits a market place. It is a technology which requires economies of scale in order to bring its costs down, and therefore it tends to be large in size – usually a 1,000 MW power plant, and often in multiples. It takes a long time to build, so capital is tied up for years without a return and without a guarantee that it will be competitive. It is an inflexible generator, which places requirements on the electricity system to ensure the generation is taken thereby excluding other generation. Nuclear power therefore requires two fundamental factors: the price of electricity has to rise until it becomes competitive (whether in the market or as a result of a secure, domestic carbon price); *and* there needs to be certainty that its output will be bought at that price for a certain amount of time.

The 'new' situation in a climate change world is the pricing of carbon. To a degree, the Government can argue that setting an appropriate carbon price would meet the needs of the political paradigm. It is carbon- rather than technology-specific; and it is mimicking a market, and therefore should provide incentives for the cheapest low carbon technologies under that price. If Government is to establish policies to meet the challenge of climate change then this is a suitable way of doing so.

However, investment in nuclear power will not occur with only setting a high enough price of carbon. It will also have to be guaranteed for long enough. Moreover, Government will also have to take care of insurance liabilities, and more crucially, capping the costs of nuclear waste to be borne by the investor. At the moment, those waste costs are unknown.

Moreover, any construction overruns would also have to be taken care of by the Government. At some point, these extra construction costs would eat into the expected returns from the carbon price and that risk, given the history of the time taken to build a nuclear power plant in the UK and, more importantly, in Finland,[6] is likely to be too great for investors. Government will have to intervene in these ways and this does not fit with the principles of the paradigm but, importantly, Government can now argue that through the price of carbon it is not picking winners or intervening in the electricity market.

Principles of public expenditure

The principles upon which Governments spend public money have been developed over a number of years, with Waldegrave's fundamental 1992 Science and Technology Report being particularly influential (House of Lords Science and Technology Select Committee, 1992). These principles have since then increasingly related to the stimulation of innovation, but continue to support the view that, as far as possible, this should occur without intervening directly in markets.

Technologies need financial support at the differing stages of their development. For example, R&D funds enable early stage development. Demonstration funds may then be required, followed by venture capital injections. All phases in the technology chain are important. The cost of technologies will come down as their volume manufacture goes up. As they become cheaper, a virtuous circle develops: their use increases, which leads to increased manufacture and a reduction in price. However, technologies often will not get the opportunity to achieve those price reductions if the early price of the technology is too high and if there is no means for it to be reduced. The UK Government has accepted that funding the various stages between demonstration and the market is an appropriate use of Government money, and this funding has risen rapidly since 2001 (UKERC database). Where possible though it establishes 'market mechanisms', such as the Renewables Obligation, to support technologies. This means mechanisms which are complementary to, or mimic, the 'market' but are external to it, so that market rules do not have to be changed to deliberately 'pick' a technology or a fuel.

A slightly different policy has been for the Government to provide support for technologies based on *developing options* and the expectation of price falls (DTI, 2003). In this situation, public support for a technology is established for a certain length of time, related to government spending periods or specific policies, for example the Renewables Obligation until 2027. This is because the point of the policy is to reduce the price of the new technology, until it is competitive with the alternatives. Thus discussion of how to end support for a technology is generally written into a policy as a 'sunset' clause. Once a technology appears to be competitive, the debate begins about whether to remove or reduce support. This is currently the case with onshore wind and landfill.

It is reasonable to expect that the same principles of support for one technology should be used to assess the type and level of support for another technology. In other words, we should expect an underlying logic to public expenditure. Yet support for a nuclear power programme does not fit with this reasoning:

- Public money would be spent on a technology which had already received a great deal of support for fifty years, and the final price per kWh is not expected to fall any more.
- That support was for an unknown amount (i.e. the total amount) for an unknown period (i.e. the number of years) without a sunset clause, locks the UK into support for a technology for an indefinite amount of time.
- This money was being given to a tried technology, despite the existence of several other technology options which appeared to offer similar or greater possibilities but which had not received support.
- Support was being given despite there being a good chance that by doing so this would undermine the other options, which are arguably the real answer to the climate change challenge.
- The support will create additional costs through radioactive wastes and decommissioning, again without trying to develop other options which do not have these problems.

The value of flexibility and resilience – as part of a successful strategy for security of any future

The value of flexibility and resilience has developed in importance over the last decade or so, particularly with respect to energy security issues. The uncertainties of the future energy system in the UK would seem to almost self-evidently preclude committing to a large new reactor programme. The inflexibility of such a large programme would be exacerbated by policy makers' susceptibility to 'entrapment' when dealing with large nuclear investments, which has been highlighted by William Walker (Walker, 2000). He discussed the long-lasting political debate surrounding the Thermal Oxide Reprocessing Plant (THORP) at

Sellafield: despite the collapse of the economic and technical justifications for operating the plant, the industry and Government continued to press ahead with plans to open it. In the event, THORP has consistently failed to operate according to BNFL's original claims, and is currently closed following an accident.

This is not just a problem experienced by one nuclear project – the original rationale for building the Sizewell B nuclear power station was that demand would grow significantly and that the station was the cheapest way of generating power to meet that demand. In the event, demand growth did not meet predictions, and the station produces more expensive power than coal, gas and even some renewables. Despite warnings of the high cost of power early in the construction project, the industry pressed ahead with building it, and the final justification for its operation was that it should be operated simply because it had been built and could be operated (Walker, 2000).

However, flexibility and resilience are widely seen as important factors in a secure energy system. To be too reliant on inflexible plant or to have limited resilience to shocks or difficulties adds to energy insecurity. Nuclear power is both inflexible and undermines resilience.

Intervention for nuclear power but not sustainable energy

Supporting nuclear power via a carbon price fits the political paradigm closer than the needs of a low carbon energy system based on renewables and demand reduction. It allows the Government to think of climate change as a technology problem; it does not require much change to the energy system and it can be implemented immediately, without making too many demands on society, even if its appreciable effect will not be for some time. However, while it may be 'easier', nuclear power cannot be the answer to forming a sustainable low carbon energy system, and may be making the situation worse. Luckily for the Government, the Stern Review's argument for setting a high domestic price of carbon has given it the means it needs to provide support for nuclear power, albeit in a non-specific manner. Setting such a carbon price avoids the need to intervene directly in the electricity market while at the same time fitting in with the other principles of the paradigm – not picking winners; and establishing a carbon reduction mechanism rather than a specific pro-renewable energy policy. This is the new and powerful development in the history of nuclear power.

In early 2007 the Government signed up to a new and radical EU renewable energy policy which requires that renewable electricity provides around 35 per cent of electricity demand in the UK (depending on burden sharing), and that total energy demand is reduced by 20 per cent, by 2020 (DTI, 2007b; REA, 2007a). This, in principle, knocks out the need for nuclear power. The Government still pressed on with its nuclear policy in the 2007 EWP despite the risks; despite the fact that this cannot be the 'answer' to the transformation to a low carbon energy system; and despite the new EU Climate and Energy Policy. This book now argues that if only the Government had put in place well-evidenced policies for renewables and demand reduction, it could have had a successful renewable energy policy which delivers substantial amounts of electricity capacity, thereby undermining the argument that nuclear power is necessary because renewable energy delivery is so poor. The Government preferred to press ahead with nuclear power because it finds the demands of renewable energy and demand reduction too difficult to deal with, given the preferences of the paradigm. This reflects its inability to deal with climate change and the wider issue of environmental sustainability and bodes very badly for UK sustainability policy until the basic principles of the paradigm change. The next chapter develops this argument in full.

5
Renewable Energy in the UK

The previous chapter explained why nuclear power has re-appeared on the UK energy policy agenda. It concluded that nuclear power suits the current political paradigm in every respect, except for three very important issues related to economics: competitiveness, 'not picking winners' and intervention. The determination not to intervene has, however, not really been tested. Nuclear power was supported with risk-free contracts and payment under the Non-Fossil Fuel Levy from 1990 until portions of the industry were privatized in 1996. British Energy, the owner/operator of the nuclear power plants, made some fundamental mistakes in relation to the electricity market, but was baled out by the Government (Woodman, 2007b). Nuclear power has been somewhat out in the wilderness since then, but fundamental support for it has never been lost. As climate change (and the need for low carbon generation) and energy security have crept up the political agenda, nuclear power has increasingly been seen as a means to meet those challenges. The recent decision to establish a high domestic price of carbon also enables a long-term price floor, or a means for a cost for difference contract, for nuclear power. In this way, nuclear power has reclaimed its position within the political paradigm. And in the same way that the re-emergence of nuclear power shows the underlying character of the paradigm, so does the relative demise of renewable energy and demand reduction.

As this chapter explains, renewable energy has never been fully supported in the UK. In a sense, to have done so would be to have moved to a 'new' or sustainable energy paradigm. It was supported initially in a risk-free manner on the back of the need to privatize nuclear power quickly in 1990. In 1996, once nuclear power was mostly privatized, the fundamental driver of a risk-free policy was removed. Renewable energy entered an era where the 'fight' was to retain a technology specific policy at all, never mind a risk-free one, given the substantial support for a broad carbon, non-technology specific policy. However, with the increasing importance of climate change and energy security, it became clear that a clear future price of carbon was required for energy investment decisions. By setting a high domestic carbon price, as called for by Stern, the Government could yet again leave energy policy decisions to the market to decide which technologies to support. To an extent, a high price of carbon will benefit renewables. But it also provides support for nuclear power. In a world of limited resources, the reality is that support for one technology means less for another. The level of commitment required to develop nuclear power is such that it must undermine other options (Mitchell and Woodman, 2006). Moreover, electricity, and nuclear power's portion of it, is only a very small part of the energy sector which needs to be de-carbonized. Undue focus on nuclear power increases the sense of political risk that the Government is not committed to dealing with the real issues of climate change but only those aspects which are (arguably) the easiest to deal with. Renewables and demand reduction are the set of technologies available to meet the wider de-carbonization requirement. The 2007 Energy White Paper has shown that the Government is not prepared to go down that path.

The extent to which renewable energy is supported for itself almost becomes the litmus test of whether the political paradigm has shifted. Energy policy in the UK has not fundamentally changed, despite a fundamental change in the drivers of energy policy. The Performance and Innovation Unit (PIU) Energy Review in 2002 was the 'furthest' energy policy got away from the current paradigm. However, resistance (Rotmans et al., 2001) kicked in and the 2003 and 2007 Energy White Papers have firmly returned energy policy to the current paradigm fold.

This chapter provides a very clear case study of how the principles of the political paradigm constrain effective policy design. It also shows how an energy policy based on renewable energy and demand reduction does not fit with the paradigm, and why the paradigm continues down the large-scale, few large companies, centralized route. It explains how the policies, rules and incentives in place for renewables can act as barriers, or spurs, to their development. This chapter explains why it is that the UK, which is probably the most vocal country in support of renewables in Europe, does so poorly in terms of renewable energy deployment, relative to other European countries. It shows that this is directly to do with the UK's political paradigm. Chapters 7 and 8 will explore the policies in place for sustainable energy in New Zealand, Germany, Spain,

Denmark and the Netherlands and highlight their differences with the British policies, and their implications.

Renewable energy in the UK

The UK has had a delivery programme for renewable electricity since 1990. Initially, the support mechanism was the Renewable Non-Fossil Fuel Obligation (NFFO) (Mitchell, 1995, 2000a) and then, since 2002, it became the Renewables Obligation (RO) (Mitchell and Connor, 2004; Mitchell et al., 2006). The UK Government has been consulting on how to amend the RO for England and Wales since 2005, while the Scottish Executive (SE, 2006) has been consulting on how to amend the Scottish RO (SRO). The latter has also been exploring options to support marine technology in particular. The 2007 Energy White Paper announced various changes to renewable energy policy, including adding 'banding' to the RO. This meant that different renewable energy technologies would receive different payments. In addition, there have been a number of research and development programmes in support of renewables. These fell to an all-time low in 1997, just prior to the incoming Labour Government, but have risen steadily since then (UKERC database).

The history of support for renewable energy in England and Wales and Scotland has been characterized by incremental change since 1990. The situation has become complex; one which almost no-one considers to be good; and one which is perceived to be less cost-effective and efficient in delivering renewables than a feed-in tariff (FIT), the type of mechanism used widely within the rest of Europe. That the RO's basic design is still supported by Government in the face of evidence of more successful delivery programmes elsewhere is in part due to the fact that it is very difficult to move from one policy to another without negatively affecting those already involved in it. There are a great many companies who have invested hugely in the RO on the basis of it continuing until 2027. Clearly, such companies should not lose out from their RO contracts. Equally, it is not responsible leadership to continue with a poor policy simply because of complaints if it is changed.

However, the current situation also stems from the Government's continued belief in the importance of maintaining an economic design (or mimicking) of mechanisms of support. The Government appears to take the view that to do so will, in some undefined way, lead to a 'better' outcome than if it had intervened more directly in support of renewables, thereby setting in train unspecified but 'unwanted' outcomes. Even more though, there is a preference for dealing with just a few 'big' companies. An energy sector made up of numerous separate companies of various size, with many technologies at various scales, dealing with a multitude of customers of different sizes and wishes, is very dissimilar to the centralized system in place today. As the DTI Minister, Alistair Darling, said at a New Statesman lunch, he finds it easier to deal with a few large companies than millions of customers (*New Statesman*, 2007).

This chapter tries to reveal the implications of this way of thinking and acting. It is not wrong for Governments to have principles; indeed all Governments have an underlying set of beliefs which make up their political framework. Nor is it wrong to support pro-market principles, since the market will no doubt continue to be the bedrock of society. However, the globe is a fast moving place whether with respect to climate change or energy security issues, and the Government has to be nimble enough to chart its way through these new problems that are testing its leadership. Flexibility, along with diversity, is an important attribute for resilience to the current and future demands made on society. So far, the UK seems unable, with respect to climate change, to be flexible.

A foot in the door – the Non-Fossil Fuel Obligation

The NFFO was primarily set up as a means to subsidize nuclear generation in 1990, which had proved too difficult to privatize (Mitchell, 1995; Surrey, 1996). The UK Government was required to ask the European Commission for permission to support nuclear power. The Government preferred to ask for an obligation to support 'non-fossil fuel' technologies, and renewable energy was included as an after-thought. The Electricity Act 1990 enabled the raising of a fossil fuel levy to pay for specified non-fossil fuel technologies, with both nuclear power and renewable energy technologies defined to be 'eligible' for such support. This beginning epitomizes the subsequent history of renewable energy support in the UK. An opportunity arose to support it, as a result of another policy demand. The justification behind the policy was never clarified or widely agreed. It was opportunistic or the equivalent to 'a foot in the door'. Once opened, the door has proved impossible to close by those who do not support a renewables specific policy.

From an economic perspective, a sector-wide carbon reduction policy – whether a carbon tax or a carbon trading scheme – is considered more economically efficient than a technology specific policy. This is based on the view that economic actors act rationally. If the ability to choose is passed to those actors, then they will make the most economically rational choices. Supporting such a policy is therefore assumed to reduce the chances of politicians or civil servants making a wrong technology choice. Those who support this view, would argue that a renewables specific policy is not only economically inefficient but is also too much like 'picking winners'. Moreover, support for a broad carbon-based policy implies that the reason for supporting renewables is fundamentally confined to carbon reduction, as opposed to wider reasons such as energy security, diversity, skills development, regional development, industrial policy, and so on.

The argument of whether to establish a renewables specific mechanism versus a general carbon mechanism has rumbled on in the UK since 1990. The Government has so far always come down in favour of a renewables specific policy, arguing that there are valuable reasons for supporting renewables other than carbon reduction. Examples include: as part of an innovation policy; to provide energy options; to support diversity; and for broader industrial and local benefits (PIU, 2002; DTI, 2003). Nevertheless, the strength of support for a sector-wide carbon reducing policy is powerful and combines different groups of support, whether it be laissez-faire economists or the pro-nuclear lobby. As a result, support for specific renewable energy policies has never been powerful or widespread across Whitehall and has never been forceful enough to push through a policy more in tune with the innovation policy arguments of risk reduction (discussed in Chapter 3) or one which will deal with the challenges of de-carbonizing the whole energy system. This division between those who support broad carbon versus technology specific policies meant that there was always a lack of clarity and agreement over the reasons for, and goals of, a renewable energy policy. This has dogged and constrained the cost, design and success of the renewable energy policy in England and Wales ever since.

Increasing, not decreasing, the risk of investment in renewable energy

An important point about the birth of the renewable energy policy is the extent to which its circumstances shaped the design of subsequent renewable energy policies; was the ability to support renewables in another way closed off as a result of its sudden inclusion in the NFFO? With hindsight, it can be seen that the pro-market approach to renewable energy support became stronger as the decade progressed. This was primarily because renewable energy was supported initially on the back of nuclear power, which needed a comprehensive, risk-free means of support. The nuclear and renewable energy generators did not have to become involved in the electricity market. The latter had to successfully bid in for a power purchase contract. Once they had that contract, it guaranteed them that their electricity was bought at the price they bid in at, for a certain period of time. This was an almost risk-free contract which enabled the projects to be financed. Moreover, the NFFO was separated into different technology bands so that one wind energy project was bidding against another wind energy project. Different technologies were not expected to bid against each other and this meant that more immature (and hence more expensive) technologies were still able to receive contracts. Renewable energy gained at this point because of the requirements of nuclear power.

However, once nuclear power had been (mostly) privatized in 1996, there was less support for such an interventionist risk-free policy for renewables and more support for a technology and fuel blind approach to technology development. The risk-free nature of the NFFO was the price which had to be paid to privatize the electricity industry, but Government was not prepared to continue this risk-free policy for renewables. The new Labour Government undertook numerous policy reviews once it gained office in 1997. Any changes to any policy, including energy (and renewables) were slow. However, by 1998 a Utilities Bill Team was established in the DTI with the intention of altering the basis of utility regulation (gas, electricity and water) in the UK. This meant that the NFFO had to be amended because the NFFO obligation was placed on the combined distribution and supply companies (the regional electricity companies) which were going to be abolished in the Utilities Act by dividing them into distribution and supply companies.

Had the NFFO been seen as a successful mechanism, it could have continued in a slightly different form. However, it had been seriously unsuccessful in delivering new capacity. Contracts were awarded as a result of a competitive bidding process in different technology bands. The developers were paid the price per kWh they had 'bid in'. It turned out that the prices bid in were too low and this meant that the projects were uneconomic to develop, with only 13 per cent of the NFFO contracts ever built (Stenzel and Frenzel, 2007). This poor result was due to competition for contracts being intense because of a pent-up demand for support for renewable energy; the very limited amount of funds available; and because of the lack of a penalty for those who had a contract but did not develop the project. Had there been more funds, competition would have been less intense. Had there also been a penalty, the prices bid in would not have been so low. Together, a higher proportion of projects would have been built.

Nevertheless as it was, the renewable NFFO was seen as an unsuccessful mechanism. Moreover, with the majority of the nuclear power sector privatized in 1996, there was less support for such a risk-free contract for renewable energy. The Utilities Act was an ideal opportunity for 'improving' the renewable energy policy or getting rid of a renewables specific policy altogether, depending on your point of view. The Utilities Act led to three major implications for renewables:

- The Regional Electricity Companies (RECs) were the legal entities on which the NFFO was placed. The Utilities Act separated the RECs into distribution and supply companies, thereby removing the legal basis of the NFFO and requiring either that the NFFO was transferred within the new legislation or that a new mechanism was put in place.
- New Electricity Trading Arrangements (NETA) were implemented in April 2001.
- The duties of the regulator, Ofgem, were slightly altered but the thrust still remained competitive: 'protecting customer interests wherever possible using competitive means'. The regulator's role with respect to the environment was marginally increased by 'having regard to' guidance from the Government and publishing an annual Environmental Action Plan.

The mechanism to replace the NFFO – from the perspective of the Government – had to counter its supposed defects, including:

- its inability to deliver deployment (as was argued above, primarily the result from a low cost-cap rather than the NFFO itself);
- providing an alternative to the must-take contracts placed on regional electricity companies (must-take contracts were thought to separate the renewable generators too much from the reality of the market place, although it was the key reason why the NFFO was perceived to be a risk-free contract for those companies lucky enough to get one); and
- that it shouldn't 'pick winners' as the NFFO technology bands were deemed to do.

The Renewables Obligation – a Labour Government mechanism

The details of the RO were finally announced at the end of 2001, with the commencement period in April 2002. This effectively reversed the rules of the NFFO (Mitchell et al., 2006) and meant that the renewables industry had had to wait four years since the last chance of a subsidized contract in the fifth and last NFFO auction in 1998. Now the industry had to negotiate an entirely new mechanism and new electricity arrangements (discussed further in the final chapter).

The Renewables Obligation was placed on suppliers to purchase and supply a certain amount of *generated electricity*. This is not an obligation to provide contract for generation from specific projects – which is an alternative, and less risky type of generation obligation (REN21, 2007; and discussed in more detail in Chapter 8). Suppliers were to source 3 per cent of their total annual supply in the period 2002–03 (initially rising to 10.4 per cent in the period 2010–11) from a list of eligible renewable electricity technologies. There was no longer a must-take contract for renewable electricity, as the NFFO was, and no price or contract length was stipulated. The renewable energy developers had to negotiate with a supplier for all aspects of the contract. Suppliers have the right to offer any type of contract that they wish. Their only obligation is to source a percentage of renewable electricity, how they do it is up to them. As is to be expected, the supplier's preference is to provide contracts to their own generator subsidiaries. They also want flexibility and would not, in preference, wish to become contracted for specific generation for too long in case they become trapped into a high priced contract. The risk involved in the RO is therefore greater for developers than it was with the NFFO. And this is because of the three types of risk inherent in the RO:

- price risk (generators do not know what they will be paid beyond the (short-term) contract);
- volume risk (generators do not know if they will be able to sell their generation in the future, certainly once the current 10 per cent target for 2010 is met);
- market risk (generation value varies according to market rules).

As was intended, the RO is far more of a market mechanism than the NFFO. It would force renewable developers to take part in the electricity market, and in this sense it has been successful. The RO is technology non-specific; all eligible generation technologies (whether landfill gas or wind energy) receive roughly the same payment, and prices are currently significantly higher than those awarded under the later rounds of the NFFO. Indeed the payment is now equivalent or higher to that currently guaranteed for a minimum of five years for wind energy in the almost risk-free EEG in Germany (Mitchell et al., 2006).

It is also a complex mechanism to administer and be involved with. To comply with the RO, suppliers have to prove to Ofgem, the energy regulator, that they have met their obligation by providing the requisite amount of Renewables Obligation Certificates (ROCs, where 1 ROC = 1 MWh). Suppliers can obtain these either directly from a generator (by buying both the energy and the ROC, or just the ROC), or by buying the ROC in a trading market. Moreover, the supplier can 'buy out' of the obligation if it does not want to participate by paying 3p/kWh for every unit of renewable electricity it should have bought to meet its obligation. This 'buy-out' revenue is then recycled back to the suppliers who have participated. A supplier submitting 5 per cent of the total ROCs submitted would receive 5 per cent of the recycled 'buy-out' or premium. The recycled 'buy-out' funds or the recycled premium, as it is known, adds a new dimension to the RO because it increases the incentive to 'game' on the part of the suppliers.

The risks involved with the RO are such that it does not provide the basis for obtaining finance. Only companies with enough corporate assets are able to take the RO risks because they are able to obtain investment capital for the RO based on their corporate assets. Some new entrants have come into the renewable energy world, although most of these were created within the NFFO. Because it is not possible to obtain finance based on the RO contract, the UK has a very high concentration of renewable development within the utility sector (Stenzel and Frenzel, 2007).

However, for those companies able to take the risk, the price paid within the RO for the renewable electricity is high. It is therefore a good mechanism for the financially strong, incumbent energy companies in the UK, and this is one reason why there is such a powerful lobby in place to continue it. The mechanism does not support new entrants, which would be threatening to the incumbents; their generator subsidiaries get paid a high price for their electricity; the extra ROC cost of the renewable generation can be passed on to customers; and they can market themselves as the chief developers of renewable energy in the UK. Thus, the RO is therefore an even stronger mechanism in support of large companies than the NFFO.

Concerns with the RO and divergence from nuclear power

There have been concerns with the RO, almost from its point of inception, and from all sides of the energy sector. Those who wanted a supportive risk-free mechanism for renewable energy complained about its design (e.g. Mitchell and Connor, 2004). But it was also criticized from the wider energy policy perspective by those who were worried about a developing 'energy gap' (CBI, 2005). The argument of this viewpoint, discussed in the previous chapter, was: if renewable deployment was poor, then a 'generation gap' would develop and renewables would not be able to fill it thereby requiring additional low carbon electricity capacity, hence the 'supposed' need for nuclear power. The implication of this view is that the problem is with renewable energy. However, this book argues, if the policy is right then renewable energy development occurs, as is shown in several countries, and discussed in Chapter 8.

The difficulties of deployment of renewable electricity in the UK are, to a very large degree, because of the RO. Of course, many other factors come into play, the most important being investment risk, grid connection and obtaining planning permission. But these parallel difficulties are either overcome or reduced with other renewable energy support mechanisms, notably the FIT.

However, as noted, the RO is a risky and complex mechanism which has a number of shortfalls. Its success, to the degree that it has any, has been with the cheapest renewable energy technologies. Other generation technology, for example wave power, tidal power, energy crops and photovoltaics, which is more expensive than the effective RO price cap has not been supported through the RO. In order to develop technologies other than (large) wind energy projects, the Government realized that it needs to bring generation from emergent technologies into the RO, or to provide support outside the RO. The Government has also made some effort to put money into overcoming the barriers to support non-electricity renewables (i.e. heat producing renewables); and to support community and small scale renewables. Renewables which are non-electricity based and which are not sold to suppliers are not eligible for the RO. They therefore do not receive ROCs and have no direct access to the recycled buy-out fund.

Stepping back from paradigm shift

It has been a very 'busy' time for the energy sector in the UK since 2002. The PIU Energy Review was published in February 2002, and exactly a year later saw the follow-up publication of the White Paper *Our Energy Future – Creating a Low Carbon Economy* (PIU, 2002; DTI, 2003). This was followed by a serious amount of lobbying and media interest and a consultation of the Renewables Obligation (2005). Altogether, this led up to another Energy Review Report (July 2006), another consultation on the RO (October 2006) and an Energy White Paper (May 2007). In parallel to this, the European Commission has also instituted a number of important policies. The first was the Renewable Energy Directive which required that each member state had a renewable energy policy in place by April 2001 and which would lead to, amongst other requirements, an additional 10 per cent of renewable electricity by 2010. Then, in March 2007 as part of the EU's Climate Change and Energy Policy, the UK signed up to a binding target for 20 per cent of the EU's energy consumption to be provided from renewable energy sources.

Despite this activity, very little has changed for renewable energy in the UK in this time. Targets for renewable electricity have doubled but rates of deployment remain poor. Nor is there any obvious force for change on the horizon. If anything, the UK has stepped back from the recommendations of the PIU Energy Review in 2002, which would have represented a (near) paradigm shift in energy policy, to one of a much more business-as-usual policy.

The 2003 Energy White Paper, published in February, set out a visionary future of a very different energy system and one that would produce 60 per cent cuts in carbon dioxide emissions by 2050 from 1990 levels. However, limited substance underlay its vision. It confirmed, just to meet the existing 2010 target of 10 per cent of electricity coming from renewables, an estimated new investment of between £1.1 billion and £1.5 billion each year would be required. To deploy renewables at the rate required to meet 60 per cent cuts by 2050 would require substantially higher investment levels, and this would in turn need confidence in the Government's intentions towards renewables. However, far from increasing confidence, the White Paper managed to increase uncertainty with respect to renewable energy policy in four key ways:

- It did not increase the target from 10 per cent of electricity from renewable sources by 2010 to 20 per cent by 2020, as was expected.
- It sets carbon trading as the centre of environmental policy, undermining confidence in the long-term existence of the renewables specific RO.
- It set up a review of the future of the RO (the current renewables delivery mechanism) in 2005/06, which in the absence of a 2020 target, raised uncertainty about change.
- It set up a review of co-firing rules within the RO, offering the potential to increase eligibility for specific technology use, thus increasing the number of ROCs likely to be generated and as a result undermining confidence in the value of ROCs.

The only positive, concrete outcome of the White Paper for renewables was an additional £60 million of capital grants over the 2002–05 spending review period. Effectively, the RO – put in place in 2002 and intended to last until 2027 – was being questioned by the 2003 White Paper, within a year of its inception. Moreover, by December 2003, concerns about the feasibility of the White Paper's renewable energy policy caused the Government to increase the obligation to 15.4 per cent by 2015. That this occurred within a year of the 'definitive' Energy Policy White Paper underlines the UK Government's seeming inability to establish long-term, workable policies.

Leading up to the 2007 Energy White Paper

In 2005, the Government consulted on the Renewables Obligation, as set out in the 2003 EWP. This brought in the idea of 'banding' technologies, meaning that different technologies would have a different ROC value per MWh. Currently, there is no difference in value of generation between technologies – each ROC equals a MWh of output from *any* technology. In principle, of course, it is good that the Government increases the payment for the less mature technologies. However, it explicitly threw out such a suggestion during the development of the RO from 1997 to 2001 because of the proposed complexity and because of the element of picking winners rather than letting the incentives lead to the appropriate technology development. Moreover, one of the reasons why an obligation was chosen over other measures was because it included the possibility for suppliers, which did not want to contract directly with a renewable developer for their generation, to buy the required ROCs from another supplier or from a broker. An important reason not to band at that time was for ease of trading. It becomes more complex to trade ROCs which have different values. Introducing the idea in the 2005 Statutory Consultation introduced uncertainty into the renewable policy. The Government's report on the Energy Review, *The Energy Challenge* (called the Energy Review Report to differentiate it from the 2002 PIU Energy Review), was released on 11 July 2006. It stated that UK Energy Policy needed to meet two challenges:

- to tackle climate change by reducing carbon dioxide emissions; and
- to deliver secure, clean energy at affordable prices, as we move to increasing dependence on imported energy.

At the same time, the Energy Review Report launched eight consultations, including one on a New Nuclear Policy Framework. It added further anxiety to the renewable energy policy debate, firstly by explicitly stating (what was, to be fair, anyway expected) that the merits of nuclear power would be re-analysed; and secondly, that it would launch *another* consultation into the renewables obligation (launched in October 2006) but raising the spectre that if any banding were to occur within the RO there should be no net gain from the RO mechanism already in place. Although this was not fully explained, the inference was that some renewable energy technologies would have their value of 1 ROC to 1 MWh of generation reduced while some would be increased. The fear was that those technologies which would find their ROC values reduced would include onshore wind.

The 2007 Energy White Paper set out its broad energy policy. With respect to renewables, there were two direct decisions:

- An increase in the Renewables Obligation to 20 per cent by 2020 on a guaranteed headroom basis – meaning that if enough renewables have been developed then the Government guarantees to keep on increasing the RO to 20 per cent.
- The introduction of banding to the RO, so that the renewable electricity from different technologies is valued differently in relation to 1 ROC per 1 MWh. The Government is consulting on this but has put forward the suggestions of reducing the value of generation from established technologies, such as sewage gas and landfill gas to 0.25 ROC/MWh; to keep onshore wind, co-firing of energy crops, and energy from waste with CHP the same at 1:1; increase it to 1.5:1 for offshore wind and dedicated regular biomass and 2:1 for wave, tidal stream, PV solar and so on.

This was reasonably good news in that onshore wind remained on a 1:1 basis. Nevertheless, the renewables industry should never have been subject to uncertainty concerning the details of its future policy.

Indirectly of course, the 2007 EWP was hugely unsettling to those thinking about investing in renewables because of the statement 'the Government's preliminary view is that it is in the public interest to give the private sector the option of investing in new nuclear power stations' (DTI, 2007b p17, executive summary). It is preliminary because Greenpeace had taken the Government to Judicial Review over the 2006 Energy Review Report arguing that the Government had inappropriately set out the intention towards nuclear power. Greenpeace won its case and the Government has been forced to consult yet again on the future of nuclear power. It is of huge import to the renewable energy industry whether the Government embarks on a nuclear power programme. Re-introducing nuclear power into the energy policy mix further undermines certainty in Government thinking and increases the attitude towards political risk, which in turn affects investment levels. Government has overturned another energy policy in only four years; who is to say this one won't be overturned in the same time?

Furthermore, the Government's reasoning was poor in the extreme. In its discussion of renewable energy policy, the EWP gave one reason for not moving to FIT as being that the NFFO was a feed-in tariff and had not been successful (DTI, 2007b p148, para 5.3.20). This is worrying. If the Government genuinely thinks the NFFO, a very different mechanism from the FIT, *is* a FIT then it is seriously under-briefed and it's not so surprising that its renewable energy policies, not to mention energy policies, richochet like a pin-ball machine. However, if it knows that the NFFO is very unlike a FIT, as it should do, then it is providing a dishonest argument. Whatever, the Government has not so far been prepared to go down the FIT route for renewables and this is fundamentally because it has not so far been prepared to intervene to this degree, whether in the design of a specific mechanism, or indeed for renewable energy. This is in marked contrast to the Government's efforts on behalf of nuclear power.

Decentralized versus centralized

The previous chapter discussed nuclear power and its complementarity to the political paradigm. This chapter is intended to explain, what is in many ways the flip-side of the argument, the paradigm's attitude to renewable energy.

A decentralized energy system with a high proportion of renewables, whether electricity or heat, appears to be only envisioned by the Government if it is linked to large companies. The Government has set in motion a number of policies related to distributed generation, including micro-generation. However, it has not initiated policies, incentives or rules to enable their development other than via the large energy companies. This complements the Government approach under the RO which supports 'big' electricity technologies (i.e. onshore and offshore wind farms) and large energy companies, because together, the Government assumes, they will provide the cheapest electricity. This is a continuation of the dominant view right from the start of the R&D programme in the UK (National Audit Office, 2005).

The Government has chosen a market-linked policy, the RO, which while expensive per kWh, has a maximum total cost each year to the consumer. The Government has made the choice in favour of the market; of certainty of what the maximum cost is; and, within some ball-park, the amount of renewable energy it will deliver. It has constrained the mechanism by requiring renewable electricity to be channelled through the utility companies thereby limiting new entrants, diversity and innovation. It has decided against a mechanism such as the FIT, which would limit risk for developers and as a result, as evidence suggests, is more successful in terms of deployment because it brings in more investment, diversity, new entrants and innovation. The Government could, if it chose to cap such a mechanism, know its total cost. The German Government did this when it first set up a feed-in tariff for photovoltaics, the so-called 1,000 roof programme.

This is discussed further in Chapter 8 but the point about the UK Government's choice is that the economic and technocratic paradigm simply finds preference for policies which appear to be economically desirable. It was always counter-argued that the effect of such risk on investment would cause the policy to be unsuccessful. This was ignored.

The Government has carried out incremental policy development, mainly to overcome problems which it was always told would occur when it put the policy in place initially.

Conclusion

The renewable energy policy in the UK is rather like a chimera. Successive Governments have always been very supportive of renewable energy in public. However, at no point has any Government ever seriously addressed renewable energy deployment, meaning that no UK Government has so far wanted to deploy renewables and then look around for a policy which would deliver that deployment. As described above, deploying renewable energy requires several factors coming into play together. All these factors are important; all have to be made 'easy' if renewable energy is to be deployed; and the type of mechanism in place can make all the difference to this 'ease'. Renewables have never really been taken seriously.

6
Markets and Networks – Pure Paradigm and Effect

A key theme throughout this book is that a sustainable energy system is one which spans electricity, heat and transport and one which encompasses all the factors that are necessary to reduce carbon emissions by at least 60 per cent by 2050. These factors include economic regulation, human behaviour, new managerial models, and so on. Another key issue is the need for innovation, and just how difficult it is to facilitate the 'right' sort of innovation. The previous chapter described the history of renewable energy in the UK and explained how the policies in place act as barriers, or spurs, to development. Earlier chapters explained how the principles of the political paradigm constrain policy design, so that policies are not as effective as they could be.

This chapter takes this argument a step further by examining the surrounding factors which renewable energy has to interact with. It shows that these areas – the electricity market and the electricity transmission and distribution networks – are regulated in such a way that they further aggravate the already poor position that the UK renewable energy industry finds itself in. Economic regulation, even more than legislated policies, is a direct reflection of the political paradigm in place. As Williamson (2000) and North (1990) explain, institutions and their principles derive directly from the paradigm. The regulators within the regulatory state are *the* embodiment of that paradigm. The combination of the policies (such as the RO or the Energy Efficiency Commitment (EEC, or the Carbon Emission Reduction Target, CERT, as it has become)) and the economic regulation and its rules and incentives go a very long way to explain why it is that the UK, which is probably the most vocal country in support of climate change reduction policies in Europe, does so poorly, particularly in terms of renewable energy deployment, relative to other European countries.

To re-cap, the UK Government has set out a number of targets to reduce carbon dioxide emissions and has established a variety of mechanisms to achieve them (DTI, 2007b). This includes a 20 per cent 'headroom' target for renewable energy by 2020; there should be 10 GW of combined heat and power (CHP) on the networks by 2010; and while there is as yet no target set for the future levels of micro-generation, it is widely discussed. Finally, the European Council has agreed to have a new EU Climate and Energy Policy of 20 per cent of total energy from renewables combined with a 10 per cent reduction in energy demand by 2020.

Some of these low carbon sources of generation will be connected directly with the transmission and distribution networks. In addition, there are new control technologies to allow the networks to be more active (or efficiently managed) thereby, amongst other things, reducing losses, and to enable customers to be more responsible for their energy consumption. Innovation is required on the part of the monopoly network operators, to allow the networks to operate more actively, and on the part of suppliers to provide new contracts with customers for services, rather than unit sales, which includes both heat and electricity. This in turn will mean that the economic regulator of the gas and electricity systems – Ofgem – will have to regulate in a way that enables such changes – or innovation – to take place. The regulator also oversees the gas and electricity markets. Their rules and incentives reflect the characteristics of the conventional non-sustainable technologies which make it harder for the new technologies to be taken up. The overall effect of economic regulation is that its rules and incentives are lagging technological change. It acts as a blocker or channeller rather than an enabler of technology.

This chapter describes how, despite efforts at stimulation, very limited amounts of innovation have occurred. It is in four sections: it examines the institutional and legal basis of Ofgem and asks whether it should be changed; it examines the electricity market and looks at whether it is appropriate, given the UK Government's sustainable development goals; it examines electricity distribution network regulation for innovation and sustainability; and, finally, it examines electricity transmission network regulation. Its conclusion is that the energy regulator is constraining the move to a sustainable energy economy and its duties will have to change, possibly in addition to wider institutional change.

Because of the process of regulation, and its interwoven nature with wider society, arguing for change to the regulator's duties is recognized as a big step. The weight of past technical, economic and policy decisions which have created the current, dominant energy system configuration tend to conspire to make such change difficult. In addition, the high degree of co-dependence of components (between the dominant generating technologies, distribution lines, transmission, markets, business and consumer behaviour both within the system and between the system and its environment) implies that the trend in any such shift would be gradual, and demonstrate incremental rather than radical innovation. This may eventually realize the policy objective, but is unlikely to be achieved quickly enough to meet the timescales set by politicians to reduce CO_2 emissions from the system. This is a fundamental problem for policy makers – they need the rapid development and deployment of both generation and network innovations to achieve their targets, but the long life of existing assets, plus the co-dependence of components means that the policy measures which will have to be set in place to achieve this must be extraordinarily far-sighted and complex. Nevertheless, the urgency of climate change and the evidence of the slow rate of change in the sphere of economic regulation means that if the Government is serious about tackling the challenges of climate change, radical change has to occur.

The actors involved in the debate about the role of Ofgem represent all energy system sectors, and often reflect polarized views which further slows change. The current incumbent energy companies argue for incremental change; the network operators often reflect very technocratic views, albeit with a very real difference between them; those of the regulator and its supporters tend to come from the economic sphere. Unsurprisingly, customers represent a wide variety of views because of their diversity, although smaller and domestic customers find it harder to get their views heard. Those who call for change tend to be those in the innovation and sustainable development world, and those who see themselves representing the interests of future customers.

This chapter tries to show, from the evidence, that economic regulation and the pursuit of sustainable development are not successfully working together. This is, unfortunately, a rather dry, technical subject. It is all too easy to switch off and think 'this is boring', but its impact is fundamental to the transition to a sustainable energy system. However, it is no mean feat to improve the situation. It is not just that setting about changing the role of the regulator is hard; it is also against the political paradigm in place. Economic regulators are representative of the paradigm! As we saw in Chapter 3, simply making any change of policy or legislation is difficult because of the internal, departmental processes.

Finally, the chapter is indebted to two sources in particular: the papers written by Bridget Woodman, as the UK Energy Research Centre Fellow on policy and regulation (Woodman, 2007a, 2007b); and discussions with the Sustainable Development Commission concerning its investigation of the role of Ofgem (SDC, 2007).

The role of Ofgem

Ofgem (the Office of Gas and Electricity Markets) is the regulator for the gas and electricity sectors. With respect to the latter, and despite its name, Ofgem has two main functions as economic regulator: the overseeing of the market for trading, and the regulation of the monopoly elements of the system. It does not directly regulate the choice of generating technology, although its design of market rules does exercise an indirect influence by the structure of trading rules and the emphasis on the lowest cost output.

Ofgem's role is defined by a set of duties which are set out in the Utilities Act 2000 and the Energy Act 2004. The duties and powers of Ofgem are endowed on the Gas and Electricity Markets Authority (GEMA) and in this context we refer to the Board of Ofgem as the Authority. The Utilities Act requires the Authority to 'protect the interests of consumers, present and future, wherever appropriate by promoting effective competition between persons engaged in ... the generation, transmission, distribution or supply of electricity ...' (Ofgem, 2006b p107). This is its primary duty; but it is also subject to various secondary, or subsidiary, duties. The Energy Act 2004 added a secondary duty on the Authority (or Ofgem) to carry out its functions in the manner which it considers will best contribute to the achievement of sustainable development.

The Authority must when carrying out its functions have regard to:

- The need to secure that, so far as it is economic to meet them, all reasonable demands in Great Britain for gas conveyed through pipes are met.
- The need to secure that all reasonable demands for electricity are met.
- The need to secure that licence holders are able to finance the activities which are the subject of obligation on them.
- The interests of individuals who are disabled or chronically sick, of pensionable age, with low incomes, or residing in rural areas.

Subject to the above, the Authority is required to carry out the functions referred to in the manner which it considers is best calculated to:

- Promote efficiency and economy on the part of those licensed under the relevant Act and efficient use of gas conveyed through pipes and electricity conveyed by distribution systems or transmission systems.
- Protect the public from dangers arising from the conveyance of gas through pipes or the use of gas conveyed through pipes and from the generation, transmission, distribution or supply of electricity.
- Contribute to the achievement of sustainable development.
- Secure a diverse and viable long-term energy supply.

In carrying out the functions referred to, the Authority must also have regard to:

- The effect on the environment of activities connected with the conveyance of gas through pipes or with the generation, transmission, distribution or supply of electricity.
- The principles under which regulatory activities should be transparent, accountable, proportionate, consistent and targeted only at cases in which action is needed and any other principles that appear to it to represent the best regulatory practice.
- Certain statutory guidance on social and environmental matters issued by the Secretary of State for Trade and Industry.

Social and Environmental Guidance

The Social and Environmental Guidance was re-issued by Government in 2004 to help Ofgem in its interpretation of social and environmental issues. It includes the following sentence:

> Where the Government wishes to implement social and environmental measures which could have significant financial implications for consumers or for regulated companies, these will be implemented by Ministers, rather than the Authority, by means of specific primary or secondary legislation.

This sentence is interpreted by Ofgem as a strong limitation on its ability to implement solutions going beyond the current interpretation of its primary duty. The term 'significant' is subject to the Authority's discretion, as is the fact that even when there is Guidance from Ministers, the Authority is only obliged to consider the relevance of the Guidance, and may take a different view from that expressed in it.

Ofgem's governance

Ofgem has its Board, the Gas and Electricity Markets Authority, which is made up of eight non-executive and four executive members, which meets once a month. The non-executive members are paid appointments, overseen by the Public Appointments Commission, and the amount of information they have to keep up with is enormous. The formal output of Ofgem's website shows how many consultations and positions they produce. In addition, the non-executive Board members tend to be on the Board for a certain reason: expertise within energy and/or business. It would be very hard for an individual to argue convincingly against a decision that Ofgem wishes to take, even within an area in which he or she is knowledgeable. In practice, this means that within the framework of duties and primary, secondary and tertiary objectives, Ofgem (and its Chief Executive) is able to exercise a significant level of discretion in weighing up issues and making decisions.

Interpreting the duties

In taking a decision, the Authority must consider how each of the secondary and tertiary duties might be relevant. So the duty to have regard to environmental and social guidance (as issued by the Secretary of State) is just one factor influencing a final decision. The influence this tertiary duty has on the decision could be considerable, but only so long as the final decision is considered to be in line with the primary duty. Therefore Ofgem's interpretation of its primary duty is of paramount importance in assessing its perceived ability, or willingness, to actively contribute to the UK's energy policy goal of cutting CO_2 emissions by 60 per cent by 2050.

One means of evaluating the relative importance of the secondary and tertiary duties against the primary duty, is to establish an economic value or additionality of an interventionist measure. This tends to be undertaken by valuing the social cost of environmental damage, which would otherwise be caused without the new measure being put in place. The Sustainable Development Commission's inquiry into the role of Ofgem with respect to climate change argues that Ofgem is systematically undervaluing carbon in its calculations and that this has the knock-on effect of leading to very few decisions that enable environmental improvements since only those with significant environmental improvements are acceptable.

Internal processes of Ofgem – the regulatory pace of a snail

There are two different processes for decisions in Ofgem:

- either through an industry code panel, for a code modification proposal to do with the gas or electricity markets; or
- through an internal Ofgem process such as the network Price Control Reviews to do with the regulated monopoly networks.

Electricity code modifications

Electricity code modifications are taken through an industry-based panel; proposals are developed and consulted on, and Ofgem then takes a view on whether the proposal should be accepted or not. Whilst this process does enable all industry to participate, smaller players are less likely to have the resources to participate fully in these multiple panels. The bias is often therefore in favour of large players' proposals due to their greater ability to influence the processes. In addition, sustainable development is not included as an objective against which any proposal can be proposed or assessed.

Within Ofgem, once the code modification has been through this process, a policy lead drafts a response, and invites comments across Ofgem on the proposal, which then requires the appropriate teams, including the (resource-constrained) social and environmental divisions if appropriate, to contribute to the draft response. If a decision is considered to be 'important' a regulatory impact assessment (RIA) will be carried out, which may require an additional consultation. The final decision will be made by the Authority, taking the policy advice and the RIA into account. Ofgem's decision, when published, would be the evidence for any judicial review, should it occur at a later date.

Network regulation

For any major project originating from Ofgem, such as a transmission or distribution price control, the decision-making process is different. The price controls operate through project teams with a project board. The project team develops the proposals and the project board comments on and guides the work before any papers are passed to the Executive and Authority for decision-making. As such, the project board has a very important role.

Sustainable development is represented in this process through the project board which will be attended from time to time by a member of the environmental and/or social policy units, depending on the issue being discussed. The project board is responsible for guiding the direction of the project to ensure that all issues are covered by the policy proposals. The project board however does not draft and propose the policies, this is done by the project team. The extent to which a project will include an environmental and social perspective is as much to do with the direction given from the project board as it is to do with the skill set of the more junior members of staff developing the policy papers, and the willingness of the project team. For sustainable development to be fully delivered, it is important for organizational expertise to be available across the organization (via the project board) as well as vertically, through all grades of staff.

In the absence of social and environmental knowledge within all Ofgem staff members, one of the most important ways in which this vertical integration occurs is through the production of regulatory impact assessments, which include social and environmental impacts. As discussed above, one means of deciding on the environmental impact is through a social cost of carbon. If that cost of carbon is too low, then the environment will be undervalued. However, assessing environmental effects through carbon only is a very narrow means of assessment, although understandable given Ofgem's role as an economic regulator. It is this narrowness which leads to Ofgem excluding non-monetary environmental effects – beneficial or otherwise – because of the complexity of establishing a methodology to incorporate non-monetary values.

This further links into its attitude to innovation. When evaluating the RIA of any particular measure, Ofgem finds it difficult to evaluate the non-monetary benefits thereby undervaluing the outcomes of certain measures, which in turn means that they may not be undertaken.

Nevertheless, this description of the general process of Ofgem underlines how diluting it is for sustainable outcomes. An open and transparent process is of course a 'good' thing, and was one of the original objectives of the regulatory state (as discussed in Chapter 2). However, the constipating effect of Ofgem's process is felt across its sphere of influence although the section below discusses, as a case study, the deleterious effect of the Ofgem process on change within the distribution networks.

Electricity markets – one step forward, one step back

England and Wales had the New Electricity Trading Arrangements (NETA) coming into operation on 27 March 2001, following Ofgem's submission of its initial proposals in July 1998. NETA has now being rolled out across Great Britain as BETTA (British Electricity Trading and Transmission Arrangements).

The aim of NETA was to act as far as is possible like a commodity market. Generators were no longer centrally dispatched but instead informed the system operator of their contracted output. Similarly, suppliers informed the system operator of how much they would buy. All contracts were submitted to central settlement 1 hour (initially 3.5 hours) ahead of the half hour dispatch period. After this 'gate closure' generators, suppliers and customers can submit offers and bids to deviate from their expected levels at specified prices into the Balancing Mechanism. The system operator can then accept or reject these to ensure the system is balanced, the quality of supply is maintained and short-term transmission constraints are dealt with. Prices in the Balancing Mechanism dictate the prices that must be paid by any generators or suppliers for any differences in their contracted position after real time Imbalance Settlement. If a generator has a shortfall in its contracted generation it must pay for that shortfall at the System Buy Price and if it exceeds it at the System Sell Price.

NETA was always a worry to small generators. Discussions concerning its creation coincided with the last NFFO Order and the development of the Renewables Obligation. It was always clear that the NFFO generators would be looked after and they continue to be paid their NFFO generation price. However, ex-NFFO (i.e. NFFO-1 and -2) generators, new projects and small generators (i.e. under 100 MW) were made subject to the new market rules, and would have to negotiate the sale of their electricity and the price to be paid against the background of the development of a new, but unknown support mechanism. This effectively put a halt to renewable energy deployment in the UK between 1998 and 2002.

Peter Hain, then Minister of Energy, asked for a review of NETA with regard to the impact of its first three months on small generators (Ofgem, 2001). Ofgem sent out 500 questionnaires in compiling the report. The 40 respondents represented 106 sites, of which 40 provided comparable year on year data. The data indicated that exports had reduced by 44 per cent for small generators on average, with independent CHP seeing the largest fall of 61 per cent. Taken together, the fall in exports and prices meant considerable impact on generator revenue; for example, the average reduction in revenue for wind power was 34.8 per cent.

A second review of the first year of NETA was published in July 2002. Again, the number of respondents was small (51 had comparable data for 2000 and 2001). Unlike the August 2001 review, Ofgem included the prices received by NFFO/SRO projects in their calculations. The contract prices of all renewables projects ranged from £33 to £77.50 per MWh, with an average of £50.76. Including the NFFO/SRO prices with other prices received via NETA, raised the average considerably (Ofgem, 2002, table 9.3), thereby implying that prices paid for renewables had risen. In fact, again, if the NFFO payments were separated from the NETA payments, prices were still much lower for renewables.

There are three key points connecting NETA with renewables:

- The mechanism, which is technology and fuel blind as implemented, will promote the status quo and dominant technologies and make it harder for immature technologies.
- A large, integrated energy company has a wide portfolio of generation which requires balancing within NETA so that difficulties of intermittent, renewable energy generation are seen as part of the extra risk of that technology and are incorporated into the overall decision to support it.
- From the perspective of independent generators, intermittent generation has more risk attached to it because of the greater difficulty of balancing individual plant output to a half hour. As a result, the price paid for intermittent generation will be discounted. This discount is likely to be greater than the real cost to the electricity system and to this degree NETA is not cost-reflective (Dale et al., 2004).

NETA is now BETTA. These market reforms increased competition and contributed to wholesale prices falling over that period (2001–05). Some of these price falls have been passed through to consumers, and in this respect Ofgem has clearly delivered on its primary duty, as it interprets it. By focussing on maintaining a downward pressure on wholesale and retail prices, Ofgem has contributed to achieving the Government's fuel poverty goals.

The Government has, by endorsing BETTA in the 2003 Energy White Paper, taken the view that not only does NETA require no fundamental changes but that it is correct to extend to Scotland. BETTA's rules maintain the status quo, because, being technology and fuel blind, it will choose the cheapest technology and it contains no mechanism to overcome any path dependencies inherent in the system. The UK is unique in Europe in that it allows no intervention in the primary market in support of sustainable energy technologies, yet no substantive intellectual debate occurred on this topic, whether in the Utilities Act or White Paper processes. Only the PIU Energy Review (2002) raised concerns over the issue of NETA and embedded generation although the underlying issue of the principles of economic regulation were not questioned.

However, the trading arrangements have created difficulties for low carbon energy generation. Predictability and flexibility of supply are rewarded and intermittent generators like wind and CHP are penalized, which increases costs for these generators and inhibits their development. Small-scale generators face a disproportionate burden from the transaction costs of participating in the balancing market. The balancing risk is often passed to the supply company so that energy from (independent) small-scale generators is often undervalued. At the outset of BETTA, it was envisaged that small-scale generators would sell their energy through consolidators who could achieve a better price through collective bargaining. By this it was meant that a small generator would sell its electricity via a consolidator which would sum the electricity from a number of small generators. The small generator would have to pay a certain amount to the consolidator but it was thought that the consolidator would be able to make a profit from the sum of those payments. However at present there is only one consolidator (Smartest-Energy) in the market, primarily because the smaller generators cannot afford to pay the consolidator and because their generation suffers a number of other transaction costs.

Ofgem oversees the market arrangements including deciding on modifications to the industry codes, as discussed above. In 2006, Ofgem approved a modification to the Balancing and Settlement Code (BSC) which changed the cash out payments (the penalty for being out of balance) to a more marginal basis. This had a disproportionate impact on intermittent generators, which Ofgem explicitly acknowledged in its RIA:

> ... if certain generation technologies are less reliable than others it is appropriate that they are exposed to the costs of managing this. They can either manage their exposure by contracting with the demand side or other more reliable generators. (Ofgem, 2006c)

This recent example demonstrates Ofgem's unwillingness to adopt a more constructive approach to low carbon generation and is indicative of the view taken by Ofgem that new innovative low carbon generation must fit within a market framework designed for large, centralized, fossil fuel power stations. An alternative approach would be to alter the electricity market and system to better recognize the unique characteristics of low carbon generation. A step towards this would be to include the achievement of sustainable development in the code objectives. This would allow the impact of code modifications on sustainability to be fully assessed as part of the code modification process.

In the recent decision on Connection and Use of System Code Modification CAP 147, Ofgem made the point that environmental costs could be included in the deliberations of code panels, under the objective relating to economic and efficient system operation. This change in position is welcome but does not appear to have any weight economically. Code modifications are taken through an industry-based panel. Proposals are developed and consulted on by the industry panel before Ofgem then takes a view on whether the proposal should be accepted or not. Whilst this process does enable all industry to participate, smaller players complained in the course of the Transmission Price Control Review (Ofgem, 2006c), that they often did not have the resources to participate fully in these multiple panels. This creates a bias in favour of large participants, which tend to be the incumbents, as they have a greater ability to participate in and influence the processes. The complexity of the codes and the modification process means that the electricity trading arrangements as currently operated are unlikely to be particularly supportive of small-scale generators.

Moreover, the move from NETA to BETTA created something known as the BETTA queue. This is a queue of wind farms waiting to connect to the transmission network. This is not discussed further here. However, it is an example of how by doing one thing, another detrimental effect occurs elsewhere, and this has been characteristic of the long history of the relationship between sustainable energy and Ofgem. The UK has around 20 GW of wind energy in the BETTA queue, *not* being connected to the onshore transmission network. The distribution networks have tried to encourage distributed generation (DG) and innovation, but to limited effect. The development of the regulation for the offshore transmission network has been slow, with no obvious direction for actors to gain confidence from. And the development of the rules of the electricity markets, have not suited the characteristics of the new generating technologies, so have further increased their risk of investment.

The debate about the importance of NETA (and then BETTA) to renewables has rumbled on since its inception. What is becoming clear is that if the various measures which are implemented as a result of NETA and BETTA modifications are taken together, intermittent and small-scale generation suffer disproportionately. This is not that the rules and incentives target them. Fuel and technology blind rules and incentives will favour either large-scale or conventional technologies. This is the 'natural' and expected result of such a principle. While 'sensible' from an economic efficiency, least-cost point of view, it is not conducive to enabling a transition from the conventional to a sustainable energy system. Ofgem, and the Government, are going to have to grasp this issue.

Distribution networks

The distribution networks carry electricity and gas from the transmission systems into our homes and businesses. In Britain there are 14 electricity distribution networks owned and operated by six different companies: Scottish and Southern Power Distribution, Scottish Power Energy Networks, United Utilities, Central Networks, Western Power Distribution and EDF Energy Networks.

Whilst the distribution networks feed electricity directly into our homes and businesses, the consumer or business buys electricity from the electricity and gas supply companies. At present the market is dominated by the 'big six' supply companies: E.ON (Powergen), Centrica (British Gas), EDF Energy, Scottish and Southern, RWE (npower) and Scottish Power. There are also some smaller suppliers such as Good Energy and Ecotricity which are serving a smaller proportion of consumers. There are also specialist suppliers serving particular markets such as businesses. However, generators, suppliers and distribution network companies often have similar parent companies – for example, E.ON owns a supplier, generation and a distribution company (Central Networks) – in order to spread the risks of operating in the electricity market. The effects of Ofgem's regulation percolates through much of this inter-woven network. Ofgem is therefore used to working with differing sections of the same companies, as they are used to dealing with Ofgem. Moreover, they will be bringing a wider strategy to any discussion aspect, because of their wider, core interests.

Ofgem regulates the profits of the distribution network operators (DNOs) through five-yearly price controls, which cap the revenue that a DNO can collect from customers, but which also act as an incentive to make efficiency gains during the five-year period of each price review. The downward pressure on costs and a relatively short regulatory horizon inherent in the price control approach mean that little emphasis is given to the longer-term strategic development of the DNOs' assets. Since the privatization of the UK electricity industry in 1990, the focus of network regulation has been on improving efficiency and reducing costs. This has largely been successful, but it has increasingly been recognized that this approach is not sustainable in the light of ageing assets and the need for increased innovation to meet Government carbon targets.

This conjunction implies that the emphasis of regulation will have to shift to one which is more supportive of investment, new technologies and new ways of operating the networks. The challenge for both Ofgem and the DNOs is how to move from the current 'passive' distribution networks to ones where active management is increasingly the norm. Passive or 'fit and forget' networks are those with one-way power flows between the transmission network to the end user and generally do not require real-time intervention to manage them. 'Fit and forget' entails both high connection costs, and high reinforcement and operating costs as the level of generation rises. Continuing this approach would ultimately limit the amount of DG that could connect in future and would also limit the extent to which the network may be able to benefit from locally connected generation (Ofgem, 2004b; DTI DGSEE, 2004). There is an increasing recognition from both Ofgem and the industry that network design and operation will have to change in order to accommodate higher levels of distributed generation in an efficient and cost-effective manner. Depending on the scale and location of new distributed generation, this will require the networks to be more intelligently managed to ensure that they stay within operational parameters and that customer security is maintained.

'Active' management is the opposite. It increases the risks of things going wrong but, on the other hand, enables them to be managed much more efficiently from the perspective of losses (carbon dioxide); resource use within infrastructure upgrades and required peak capacity, if done properly. The key short-term concern is ensuring that technologies employed in renewals, upgrades or new connections will be capable of performing in a more active environment while maintaining network security. As Botting points out, 'The future migration plan begins now. Every piece of new equipment placed on the power network will become either a problem or part of a solution to the future architectural issues' (Botting, 2005 p17).

It has been estimated that as much as 10 GW of distributed generation will be connected to the electricity distribution network by 2010 (Ofgem, 2004b; Mott MacDonald and BPI, 2004a), and this is additional to the UK's 20 per cent renewable electricity target by 2020 and the proposed EU's 20 per cent renewable energy target by 2020. Although there has been an increase in the levels of DG since the privatization of the electricity industry in 1990, the system remains overwhelmingly centralized, with most generation being large-scale plant connected to transmission lines and electricity crossing the networks to consumers. Connecting new, small-scale generation (relative to the 400–1,000 MW centralized power plants of fossil and nuclear generation), which often operates intermittently, implies that the DNOs, and the networks themselves, will have to become more active participants in the electricity system. The problem is that although this issue has been looked at in detail since 2000, there has been very little *actual* change in the design and operation of the networks.

The mechanisms put in place via the economic regulator to promote innovation and investment in this area have had only limited success.

Ofgem and innovation in distribution networks

Ofgem's measures to encourage more innovation and DG concentrate on two areas – the process of innovation in the firm (the Innovation Funding Incentive), and the process of deployment in the system (Registered Power Zones and to a lesser extent the DG incentive). Innovation activities tend to be path dependent, both because of the established practices and expertise within companies and because the established characteristics guide assessments of what is and is not likely to become a successful innovation. Moreover, in an established system, incremental innovations are more likely to be deployed in the system because of the interests of established companies and the need to conform to dominant technical standards and system competencies. Radical innovations which potentially challenge the established characteristics of the system are, as argued earlier, less likely to be supported by the conditions in the environment in which the system operates, called the 'selection environment' (Nelson and Winter, 1977). Innovation faces a range of barriers – institutional, social, technical, as well as the 'soft determinism' of economics. The importance of the conditions in the selection environment to the success or failure of innovations has been recognized by the DTI/Ofgem group charged with ensuring that there is a strategic approach to developing future network solutions in the UK:

> The role played by the non-technical aspects, regulatory, health & safety, environmental, commercial, etc cannot be over stated. These elements will in most cases dictate the success or failure of any given 'innovative technical solution' to either be adopted or commercially viable. (Botting, 2005 p17)

The combination of increased levels of low or relatively low carbon DG, and the integration of electricity and heat networks could lead to considerable savings in CO_2 emissions. At the same time as interest in DG is growing on environmental grounds, the electricity industry is facing the need to replace ageing generating plants and upgrade or replace ageing transmission and distribution infrastructure (PIU, 2002; Ofgem, 2004b). Around 70 per cent of the UK's network assets are reaching the end of their design lives, and Ofgem is concerned that they should be replaced or upgraded on a cost-effective basis. In addition, a number of infrastructure failures, either in the UK or abroad, have heightened awareness of the fragility of networks, and focussed policy-makers' minds on the need to ensure that they are maintained or renewed in a timely way to avoid costly failures (Helm, 2005; Energy Networks Association, 2006).

The need for infrastructure renewal should provide an opportunity for the operators to begin a shift to more active management of their networks. Increasing levels of DG and active network management could reduce network reinforcement costs; Strbac and Jenkins (2001) put this at from around £50–60/kW to about £20/kW on low voltage lines in rural areas and by up to £100/kW in urban areas for actively managing fault levels. Reducing carbon dioxide emissions will entail a degree of innovation in both the design and management of the networks to accommodate the two-way power flows and the new generation. As privately owned, monopoly companies whose operations are closely regulated by the economic regulator Ofgem, the DNOs have not placed innovation high on the agenda. If there is to be a shift in how they invest in and operate their networks, this will have to be stimulated and approved by Ofgem.

Because of the diverse range of actors, technologies and potential outcomes, achieving a shift to a lower carbon, more decentralized system will require a strategic approach in the design and implementation of policy and regulation. It will also require the risks of implementing new technologies to be balanced between network operators, generators and consumers. This means that Ofgem will have to devise a twin approach to the regulatory structure where there are:

- sufficient incentives for DNOs to invest in new technologies and practices; and
- efforts to remove barriers to make investing in DG sufficiently attractive.

It also has to ensure, as far as possible, that the interests of DNOs and developers coincide – in other words, that new projects can be sited in areas where they are beneficial to both developers and DNOs.

Big effort – little change

Ofgem has been involved in analysing this area since 2000 and has concentrated a considerable amount of effort to encourage innovation. The process began with the Embedded Generation Working Group, set up by the DTI, Defra and Ofgem. Various other working groups have followed, all with similar but different names, culminating with new incentives in the 2005 Distribution Price Control. The results, unfortunately, have not been good and are a case study in why it is difficult to imagine how the type of changes required to deliver a sustainable energy system can occur at the rate required.

The price control review resulted in three measures designed to encourage innovation and greater levels of distributed generation: the Innovation Funding Incentive (IFI), the DG incentive, and Registered Power Zones (RPZs). The measures were originally put in place for the five-year duration of the price control. The general definitions for the IFIs and RPZs were set out by Ofgem, with more detailed definitions and processes for approval set out in the 'Good Practice Guide', produced by the DNOs and approved by Ofgem (Energy Networks Association, 2005).

The idea of the IFIs was to increase the incentives for DNOs to spend more money on research and development projects designed to enhance network design and operation. The incentive covers any aspect of distribution system asset management from design to decommissioning as long as the focus is on providing value for end customers by enhancing efficiency in operating costs and capital expenditure. There are various rules: approval has to be given by Ofgem; and the benefits are assessed on financial grounds, in other words, the extent to which a project will deliver financial benefits to consumers. Although both Ofgem and the Good Practice Guide include the non-financial (social and environmental) benefits of a project, these are expressed in financial terms, and the overall emphasis of the mechanism is therefore on relatively certain, quantifiable benefits from any project.

The distributed generation incentive mechanism is designed to encourage DNOs to 'invest efficiently and economically in the provision of DG connections and to be generally proactive in responding to connection requests' (Ofgem, 2005 p5) and is designed to address the connection charging issue identified as a barrier to new DG as early as 2001. Incentives are given to the DNOs whereby they no longer 'lose' money by connections (Ofgem, 2004b).

The DG incentive is not designed to encourage innovation in connections, and the returns it will bring are not intended to balance the risks of innovative technologies and connections. These risks are instead meant to be addressed directly in the commercial sphere that the DNO negotiates with the developer (Ofgem, 2004b). It should be beneficial for two reasons. Firstly, reducing the upfront connection costs for a project should in theory reduce one of the barriers for smaller generators, although, over the lifetime of the project, the mechanism will in fact lead to higher costs for distributed generators (Ofgem, 2004a). But secondly, and perhaps more importantly, further incentives are given to the DNOs to ensure that they do not lose financially by connecting DG, thereby removing the very important incentive DNOs had against connecting DG. The incentive is also not designed to encourage a more strategic view of network development: each application for connection will be assessed by the DNO on a case by case basis, rather than as part of a potential group of future projects.

The Registered Power Zones are potentially the most innovative mechanism established by Ofgem to encourage DG in that it shows a more holistic approach to overall system development than is necessarily implied by the other two mechanisms. It is aimed both at encouraging the early deployment of new network technologies or innovations, and at encouraging the connection of new generation, for example by installing a device that could improve the monitoring or control of the network. RPZs are a sector of a distribution network (either geographic or defined by electrical connections) in which the DNO can demonstrate innovative solutions to the connection of new distributed generation; with the incentive even higher than, and additional to, the DG incentive.

Because of the uncertainty attached to the performance of any innovative technology or practice, it is impossible to definitively value the three measures. Mott MacDonald was commissioned by Ofgem to produce an estimate as part of the development of the price control which showed a clear positive benefit in implementing the IFI and RPZ schemes (Mott MacDonald and BPI, 2004a). This was a necessary step for Ofgem in order for it to argue that it was protecting the interests of customers in its acceptance of extra incentives to the companies. This highlights the difficulties Ofgem perceives itself to have for stimulating innovation.

The initial design of the IFI and RPZ schemes was cautious and conservative in terms of balancing the costs of DNO activities and the possible resulting benefits. This indicates that the weight given by Ofgem to its sustainable development duty was slight in comparison to the weight given to its duty to protecting the interests of consumers and that it has not adopted a long-term view of network development and change. So far, progress with the three mechanisms has been mixed:

- The introduction of IFIs does appear to have stimulated interest in R&D and innovation in some DNOs, although others are still spending less than 0.1 per cent of their turnover. The emphasis across the DNOs appears to be on extending the life of existing assets, rather than on projects designed to enable a shift to more active network management.

- There is no increase in interest in new connections – whether rated by size or number of offers – as a result of the new 'shallowish' charging regime. This indicates that, while the DG incentive may be suitable for DNOs, it is less attractive for developers than the earlier, 'deep' charging arrangements.
- Only three RPZs have so far been approved. The overwhelming reason for this identified by DNOs was the lack of common interests between developers and network operators about siting new projects and areas where the network could benefit from RPZ status.

As discussed in previous chapters, innovation does not occur in some linear and predictable manner. It is more the case that an environment conducive for change has to be encouraged. A programme carefully designed to balance 'outcomes' with 'costs' and based on a cost-benefit analysis might stimulate innovation, but, based on the evidence we know of how innovation does occur, would seem unlikely. The IFI and RPZ are important. Partly, this is because there may be additional levels of DG. But, at this time, their primary importance is to highlight just how poor Ofgem is at stimulating change. And this can be clearly linked to Ofgem's duties, which in turn can be clearly linked to the political paradigm in place.

The transmission network

Offshore networks

Offshore wind is expected to play an increasingly prominent role in supplying renewable power in the UK. Connecting offshore renewables to the UK grid will require the construction of new networks to bring the power ashore and new commercial arrangements for generators and transmission operators. However, the process of designing a regulatory regime to encourage investment in new networks and establish a revenue stream for their operators is proving to be a complex, protracted business, and highlights the possible conflicts between an overwhelming preference for market-based approaches and the need for fast, effective action to reduce carbon dioxide emissions.

So far, the Government has conducted two rounds of bidding for offshore wind sites leases. The first, in December 2000, allocated 18 sites within 12 kilometres of the shore. Each site was limited to a maximum of 30 turbines grouped within an area of 10 km. These limitations reflected the position that the initial phase of offshore wind development in the UK was effectively a demonstration of the technology. A second round of leasing took place in 2003 to bid for sites in three strategic areas – the

Thames Estuary, the Greater Wash, and off the coast of North Wales/the North West of England. This allocated 15 sites with a potential combined capacity of 5.4–7.2 GW. Unlike Round 1, the sites are not limited to specific numbers of turbines and can be sited further from shore. Given the size of the resource and the proposed projects, offshore wind is expected to contribute a significant proportion of the UK's renewable generation by 2010 and beyond.

It was originally intended that most Round 1 projects would begin generating in the summer of 2005, and that Round 2 projects would come on line from 2007 (DTI, 2002). However, so far only four of the original 18 Round 1 projects are operating. No Round 2 projects are yet under construction, although several have received consent from the DTI (BWEA website, 2007).

The development of renewable energy technologies in the UK has been a slow process, despite significant resources. As we have seen, this has largely been because of problems within the planning system which has led to projects being delayed or rejected, and also because of the competitive nature of the Renewables Obligation. In addition to these generic problems facing new technologies, however, offshore wind projects face additional issues which affect both the progress of a project and its economics. These include:

- increase in turbine costs;
- weather dependent construction schedules;
- technical uncertainties;
- high transmission costs.

The construction of projects off the shores of the UK raised a range of new legal and regulatory issues. For Round 2 projects, these include the need to develop projects outside the UK's 12 mile territorial limit. In 2002, the Government responded to these new issues by producing a strategy for the long-term development of the UK's offshore renewable resources. *Future Offshore* (DTI, 2002) set out the Government's approach to the legal framework for offshore development which was later embodied in the Energy Act 2004. It also clearly identified the need for the efficient provision of infrastructure to allow the connection of offshore projects to the transmission or distribution networks.

Constructing new transmission lines is a significant expansion of an electricity system which has been stable in terms of its scope for decades. It also raises the possibility of a new layer of contracts, security standards and ownership. Round 1 projects are, or will be, connected to the onshore system through distribution network voltage lines. Most Round 2 projects will use offshore 132 kV transmission cables to connect to onshore distribution networks (Ofgem, 2007c). This therefore implies a three-stage connection between the wind turbine and the transmission network: from the individual turbine to the 132 kV offshore network, then to the onshore distribution network and finally the connection to the transmission network. This in turn implies that there can be up to three different owners, each of which will be required to co-ordinate their construction and operation plans through, for example, technical standards and commercial contracts. In addition to this complexity, the construction of the new transmission network in hostile offshore conditions will undoubtedly be more risky and complicated than the construction of new networks onshore.

The design of regulation for offshore transmission will therefore have to address a number of issues (Davies and Ward, 2005):

- ensuring legal continuity between the onshore and offshore regulatory regimes;
- developing a cost-effective approach to connections, when they are both constructed and operated;
- ensuring that generators with connections were not in a position to abuse their monopoly power;
- reducing the risk of stranded assets;
- reducing technical and security risks for the system operator.

In the interests of rapid development to meet the 2010 renewables generation target, as well as contributing to reduced carbon dioxide emissions overall, the regulatory regime should also be as simple and as workable as possible, and should be designed to reduce the risks of deploying a new technology. In the longer term, the regime should enable the offshore networks to be extended to incorporate other projects, whether offshore wind, or wave and tidal projects, in as effective a way as possible.

The development of the offshore transmission regime is proving to be protracted (see Table 6.1). Round 2 licences were allocated in 2003, and the Energy Act passed in 2004. This established a broad framework for the development of an offshore generating industry. It requires offshore generators, distribution and transmission operators in the Renewable Energy Zone[1] to be licensed, and gives the Secretary of State power to create and introduce a new regulatory regime for offshore distribution and transmission. Following this, there have been two major consultations on details of regulatory design and implementation, with at least three more consultations scheduled. The final Offshore Electricity Transmission Regime is intended to come into force in mid 2008 (DTI, 2006c), six years after the allocation of Round 2 sites and the strategy in *Future Offshore*.

The Government's initial position, set out in *Future Offshore*, was that there were no particular grounds for extending the onshore regulatory regime to include offshore assets, and that a non-regulated approach to construction and operation of the offshore networks could be preferable:

> The Government's conclusion is that, although extending the licences of the TSOs [transmission system operators] and DNOs offshore might be a workable solution, there are no compelling reasons for adopting this approach, rather than leaving the responsibility for providing infrastructure with offshore generators and third party providers. In particular, it is not clear that the regulated businesses of the transmission and distribution companies would have better incentives to invest efficiently in new cables than non-regulated businesses. (DTI, 2002 p72)

A non-regulated approach was endorsed by Ofgem, which envisaged a model whereby a company (including a generator) would apply for a licence to construct transmission assets, although the conditions of the licence would be limited to areas such as the obligation to offer surplus capacity to third parties and addressing issues raised by the need to interface with the onshore transmission system:

> Ofgem's initial view is that there are a number of arguments that suggest there would be significant benefits in the regulation of offshore transmission on a merchant basis and that this approach would be consistent with Ofgem's statutory objectives. (DTI, 2002 p72)

Generators would meet all the costs of developing the transmission assets. Instead of a regulated revenue stream to finance the costs of the assets, the owner would have to negotiate with generators to determine the costs for using the connection, as well as other issues such as the construction timescale and security standards to be applied. Although there may be some regulatory oversight in terms of the charges to be paid by developers, Ofgem would seek to ensure that this was as light in touch as possible. This approach would effectively put the majority of the risk of constructing both the offshore windfarm and

Table 6.1 Discussion and consultations concerning offshore transmission since 2002

	Area	Status	Date	Notes
DTI	Future Offshore	Consultation	Nov 2002	Strategy for the long-term development of offshore networks
DTI	Round 2		July 2003	Allocation of offshore licences
DTI	Energy Act 2004		July 2004	Secretary of State has the power to modify transmission licences and extend the GB system operator licence offshore
DTI/ Ofgem	Regulation of Offshore Transmission	Consultation	July 2005	Proposed two regulatory options: • price control with or without capping or cross-subsidy • licensed merchant approach
DTI	Regulation of Offshore Transmission	Decision	Mar 2006	Extension of the onshore price control regime: • Transmission Operators (TOs) to be responsible for construction and costs • Single system operator for GB • Developers pay TNUoS charges • Cost reflective to reflect demands placed on system
DTI	Offshore system operator	Consultation	May 2006	Letter announcing that the Secretary of State was 'minded to' extend the role of the GBSO offshore
Ofgem	Scoping Document	Consultation	Apr 2006	**Exclusive licenses** • Single TO with responsibility for a defined geographic area • Competitive tender approach to awarding monopoly zones **Non-exclusive licences** • TOs licences issued to any party which applies and meets the application criteria • TOs bid revenue stream • Winning TO gets price control and has assets absorbed into its RAB
DTI	Offshore system operator	Decision	Aug 2006	Confirmation of National Grid as offshore TO designate
DTI/ Ofgem	A Security Standard for Offshore Transmission Networks	Consultation	Dec 2006	Offshore security standards
Ofgem	2nd Scoping Document	Consultation	Mar 2007	Offshore Electricity Transmission Regime
DTI	A Security Standard for Offshore Transmission Networks	Decision	Apr 2007	Different standards for onshore and offshore transmission

Source: Woodman (2007c).

the transmission project on to the generating developer. The risks of developing new technologies, plus bearing the risks of the connection to the onshore network, put a huge burden on developers and the economic performance of their projects.

By July 2006, the Government decided that a regulated price control approach was in fact preferable to a merchant approach, in the interests of consistency with onshore arrangements, and to encourage a more co-ordinated approach to developing offshore networks (DTI, 2006c). It would also reduce the level of risk for offshore developers as a price control approach would allow generators to pay connection costs through annual transmission charges over the lifetime of a project, rather than upfront. Overall, the Government believed that a price control approach will result in a higher investment in offshore renewables in the medium to long term compared to the other approaches:

> Without a specific comprehensive regulatory regime which introduces an element of certainty of outcome, there is a real risk that renewable energy developers will not be encouraged to come forward with proposals for projects outside territorial waters. (DTI, 2006a para 1.12)

The replacement for the merchant approach is tendering, where different companies bid to construct and maintain the new offshore transmission lines. A tender approach, however, is also problematic. A different bidder might be successful in the three strategic areas, offering different terms to developers in each area. The bids will be based on generating licences already awarded (i.e. the Round 2 licences), with little weight given to the potential long-term development of the network to incorporate new projects within or outside the areas. The risk is therefore that tendering will lead to a modular approach to transmission construction, which is adequate in the short term but potentially limiting to the strategic exploitation of offshore resources in the longer term.

The story of the development of the regulatory regime for offshore networks so far highlights two key issues. Firstly, the drive to devise a market-based, unregulated approach led both the Government and Ofgem to endorse a merchant basis for the new transmission lines. This may or may not be suitable for conventional transmission lines connecting conventional generation, but not for new technologies being deployed in a hostile maritime environment. The shift in approach to extending the onshore price control regulation for the transmission operator's revenue was a recognition that the merchant method would be too risky. The alternative arrangements announced for the construction and operation through tendering may result in a limited potential for the efficient expansion of the offshore networks in future, while not necessarily guaranteeing that the construction and operation of the transmission lines would be achieved at much lower cost than if the responsibility had been given to the National Grid at the offset.

A second issue is the time that it has taken to reach the conclusions put in place so far. The final regulatory regime will not be in place before the middle of 2008 at the earliest, six years after generation licences were granted. Given the urgency of the need to reduce carbon dioxide emissions, and the possible contribution that offshore generation can make to this, it would have been simpler, quicker and ultimately more far-sighted to simply extend both onshore price control regulation and the National Grid as the builder and operator from the beginning. The search for cost-effective, market-based solutions to incorporating new technologies into the UK grid is effectively acting as both a barrier and a delay to the deployment of new low carbon technologies.

Onshore networks

In tandem with the introduction of BETTA, new rules and incentives for transmission networks have had serious negative effects for renewable power plants wishing to connect to them (DTI DGSEE, 2007a). The new economic framework put in place by Ofgem has resulted in a giant's step back for renewable electricity access to transmission networks. This is another complex area (only very briefly highlighted here) which illuminates how ill-suited economic regulation is for implementing economically efficient rules and incentives while at the same time enabling the development of new technologies. A generator can only inject electricity into the transmission network if they have transmission entry capacity (TEC). TEC is not tradable and was awarded to those generators already generating at the time of the move to BETTA. More TEC will be given out when there is need for more generation capacity. Until then, and this may be up to twenty years hence for some power plants, renewable energy generators cannot connect. This clearly has 'mega' impacts on the move to a sustainable energy system and the ability of the Government to meet its targets.

Conclusion

This chapter has tried to show how poor the energy regulator Ofgem has been in relation to supporting a sustainable energy system. Ofgem would argue that this is not in its remit, and to a very large extent this is true.

This therefore means that the regulator, which has so much influence on the energy system, is not supporting the move to a sustainable energy system.

However, Ofgem has a great deal of leeway in its interpretation of its duties. In this it has also been very cautious, again supporting the idea that it is not anxious to take up the challenge of delivering a sustainable energy system.

The key issue to address is whether the role of Ofgem is to transform the current energy system into a sustainable one, at least cost (defined in some long-term sense), or whether it is to maintain the system as a least-cost technology and fuel blind system, while doing what it can for sustainability. If it continues in the latter role, there will be little movement forward. We have already seen how, even when it tries, Ofgem has been poor at changing the system.

The process of change is very slow, as shown both for distribution and transmission networks. Moreover, when faced with recent decisions, after it had the sustainable development clause in its objectives, Ofgem still accepted modifications to transmission charging which further penalized intermittents. New changes, such as the implementation of BETTA, create new and unforeseen (or ignored) difficulties, such as the BETTA queue. It is one step forward, and one step back.

This book argues that Ofgem should be turning the current energy system into a sustainable energy system. However, in the view of the author, the momentum for the current system is too great and the rate of change away from it is too slow. If Ofgem continues as it is, the Government is faced with three situations:

(1) It either continues as it does, supporting R&D and deployment fairly minimally and with progress continuing to be slow.
(2) Or, it can keep Ofgem as it is but increase or change its own policies through legislation separately from Ofgem but which Ofgem then has to fit in with.
(3) Or, it acts to change the duties of Ofgem.

Unless either, or both, of (2) and (3) occur, Ofgem will continue to take the least-cost economic way forward, as its duties imply. While least-cost, these decisions tend not to be supportive of innovation or change. In order to deliver this change, rules and incentives will have to be put in place which increase certainty and reduce risk; enable new entrants; and importantly, allow actors to make money out of doing something new.

7
New Zealand as a Case Study

The intention of this case study is to illuminate how all countries have a political paradigm and that this acts as a shaper, a constrainer, a channeller or an enabler of policy development. New Zealand has a very particular character and is very independent. Its underlying political paradigm is very similar to the UK's, even though there are quite a few differences between the two. Unlike the UK, New Zealand is not a country of rigorous separation of public and private. While its energy industries work within technology and fuel blind markets, there is no economic regulator as in the UK and many of the main energy companies are still major-owned by the NZ Government. Nevertheless, the ethos of the country is still that decisions should be made through the market place; that economic analysis should be a building block of policy; that there is a limited need for innovation and that innovation which is required can be set in play through linear and predictable policies. The net effect of this paradigm is a disjuncture between the vocal pro-sustainability announcements of the Prime Minister and the policies put in place, and to this degree, the two countries are very similar.

Context

New Zealand has signed up to the Kyoto Protocol and has legal requirements to reduce its greenhouse gas emissions back to 1990 levels by 2012, or take responsibility for buying emission credits to cover the deficit. Its total emission levels are currently about 25 per cent above what they need to be to meet that commitment; 50 per cent of them derive from the intractable, and very economically important, sector of agriculture; energy demand is increasing at a historical rate of 2 per cent per annum, with the majority of new electricity capacity being fossil based; and transport emissions are also rising rapidly, in line with the rest of the globe. In order to fulfil its Kyoto commitments, New Zealand is examining its climate change policy, its energy policy and its land use policy (NZES, 2006; NZEECS, 2006; MAF, 2006).

New Zealand is similar to most other countries in that there are many different ministries and agencies involved in the development of energy policy. The Energy Minister is a junior minister (ranked 18th of 20 in Cabinet) in charge of part of the work of the Ministry for Economic Development (MED). He is also the Minister responsible for Climate Change Issues (which ought to be of benefit) but this is with the Ministry for the Environment (MfE). One can only imagine the divided loyalties and opportunities for difficulties with this mix of tasks. MED has responsibility for energy policy, while MfE has responsibility for climate change policy. Renewable energy and energy efficiency measures rest with the Energy Efficiency and Conservation Authority (EECA), which has responsibility to a Government Spokesperson on Renewable Energy and Energy Efficiency. A Government Spokesperson does not have ministerial authority and is responsible to the Minister of Energy and MED. On the other hand, EECA sits within the ministerial oversight of the MfE, which could cause conflicts (for example, what the Government Spokesperson favours may not be what the Ministry for Energy (MED) favours, or vice versa; and then there is MfE to consider). The Treasury, the Ministry of Transport, the Ministry of Agriculture and Forestry, the Department of Conservation, the Ministry for Research, Science and Technology, and the Electricity Commission are all involved to lesser or greater degree in the development of climate change policy, or other policy areas which in turn affect energy policy.

Within the climate change debate in New Zealand, there are two central issues which stand out to the 'independent bystander'. The first is how New Zealand defines itself as a 'Good International Citizen' with respect to climate change; and the second is how the NZ Government is determined to not put itself in a position where it may be backed into a corner with an open and transparent commitment to reduce carbon by a certain amount by a certain date. The interrelationship of the two issues reflects all the difficulties that New Zealand faces when developing its policies. On the one hand, being a Good International Citizen, even in the very 'weak' way of meeting its Kyoto commitments, will require both major policy action and the purchase of carbon credits internationally, both of which are significant challenges for New Zealand. Defining Good International Citizenship in some 'strong', clean and green, 'planetary carrying capacity' way in line with European policy, is nowhere in sight in practical policy terms. This is despite Prime Minister Clark's stated aspiration for New Zealand to lead the world in becoming a sustainable economy (Clark, 2007a, 2007b).

As the NZ Government knows well, if it were more open about the timetable for greenhouse gas reduction, then clearer pathways, policies and targets to achieve this would be required. For a variety of reasons outlined below, the Government is deeply divided about what these policies should be and, so far, has chosen to take a cautious path and avoid committing to anything too concrete, at least at a high level.

The NZ Government has embarked on a wide-ranging review of policies which affect the environment in the broadest sense. With respect to energy policy, a Draft New Zealand Energy Strategy (NZES) and a Draft New Zealand Energy Efficiency and Conservation Strategy (NZEECS) were announced in December 2006. The Draft NZES has six principles, one of which is 'maximizing the proportion of energy that comes from our abundant renewable energy resources'. Unlike the NZES, the NZEECS has to fulfil certain legislative requirements. The Energy Efficiency and Conservation Act 2000 states that 'the strategy must state targets to achieve those policies and objectives, being targets that are measurable, reasonable, practicable, and considered appropriate by the Minister'. The Draft NZEECS, which is part of the NZES, includes an objective of more energy from renewable sources and states that 'The Government is giving further consideration to the level and type of renewable energy targets, including whether it is desirable to establish interim milestones' (NZEECS, 2006 p49).

NZ electricity situation

New Zealand is a country roughly the size of Great Britain made up of two large islands (North and South Island), but with a relatively small population of 4 million people. The majority of those (3 million) live in the North Island, with 1.4 million living in Auckland, the business capital which is situated towards the north of the North Island. This spatial configuration of population underlies particular issues and concerns about security of supply. Electricity has historically been generated from large hydro power plants and this is linked to the 'clean and green' image which New Zealand is so proud of. The availability of hydro electricity has led electricity to be far more widely used for space and water heating than in the UK. However, since about 1980 new electricity demand has been met by new fossil-based generation. New Zealand has cheap electricity compared to most developed countries because the investment costs of the hydro power plants have long since been paid off and because the Government still has majority ownership of most of what was the nationalized energy industry through state owned enterprises. Maintaining these low electricity prices (or at least only having moderate energy price increases as a result of climate change policies) is another politically important underlying requirement of energy policy.

Despite New Zealand's energy demand still being relatively small in total; having hydro dominated electricity generation; and New Zealand having one of the best renewable energy resources in the world, it does face a number of difficult energy and climate change policy issues. The proportion of electricity generated from renewables (mainly large hydro power plants) is falling relative to the electricity generated from fossil fuels, and is now responsible for around 70 per cent of supply. Electricity demand is increasing, although MED figures project its trajectory to fall below the historical rate of 2 per cent per annum. Carbon dioxide emissions are rising steeply, because the new power plant capacity to meet this demand is primarily fossil fuel (mainly gas) based. However, CO_2 emissions should halve by 2012 in order to return them to the 1990 level, the basis of New Zealand's Kyoto commitment.

Although agriculture is the source of 50 per cent of New Zealand's greenhouse gases, there is limited agreement as to how they should be reduced. In the short term at least, the burden for change in a carbon constrained world will be from the electricity sector; and even more so from the 30 per cent of electricity capacity which is generated from fossil fuels. Carbon capture and storage (CCS) offers the hope of less dramatic change to the characteristics of the electricity system. However, even if the economic and technological uncertainties are overcome, and this is by no means sure, CCS is unlikely to be online by, or provide much carbon-free generation before, 2025.

In the short term, New Zealand may meet its Kyoto commitments by a number of means, including buying carbon permits at the global cost of carbon; putting in place a substantial long-term policy or transitional measures – the latter being the focus of another NZES consultation document. New Zealand has to switch the investment incentives from fossil fuels to renewables *before* another generation of fossil fuel power plants are built, if it is to bring carbon levels back to 1990 levels before 2030. This is almost unacceptably poor from a European public policy perspective which has Kyoto commitments to reduce, on average, their 1990 carbon emissions by 12.5 per cent by 2010, and is certainly nowhere near a meaningful definition of a Good International Citizen. It seems unlikely that, a NZ$25 (US$12–16 approx.) carbon price (whether a tax or through trading) is enough to stabilize carbon by 2030. This means that New Zealand either has to crank up the price of carbon domestically, or has to put in place some domestic carbon-reducing mechanism *in addition* to carbon trading or a tax, if it is to reduce carbon by 2030. It is important to be clear that 'reducing carbon' does *not* meet New Zealand's Kyoto commitments but simply gets carbon back to current levels. Moreover, because New Zealand has such a small electricity system, new fossil investment in the short term (for example, a 400 MW gas power plant) can alter the carbon trajectories for the longer term considerably and it is this which injects such urgency into the NZ debate – an urgency which so far appears to have been ignored by many of the players. The country cannot afford to add new non-renewable electricity generation, a point stated in the Draft NZES.

In order to stabilize carbon emissions from the electricity sector; to halt the decline in the ratio of renewables to fossil fuels and to reverse the trend in the proportion of electricity provided by renewables, a target of around 75 per cent of electricity from renewable energy resources by 2030 would be required. This means that about 200 MW[1] of new renewable electricity capacity is required every year from 2012 to 2030, which is a little more annually than the current total of 170 MW of wind energy that is currently in place. Two hundred MW of new renewables a year would fill the projected 1.5 per cent electricity demand growth and stabilize carbon dioxide emissions around the current (i.e. 2006) 5.6 million tonnes ($mtCO_2$). If renewable energy is installed below this rate, CO_2 emissions are likely to rise because the energy demand will be filled by fossil fuels, without CCS at least until 2025. Moreover, a 75 per cent penetration of renewables by 2030 would not stabilize CO_2 emissions if the annual rate of electricity demand is above 1.5 per cent, and this is quite likely given the historical figure of 2 per cent.

Thus, this is a 'crunch' time for renewable energy versus fossil fuel investment in New Zealand. Currently, 400 MW of new fossil generation is under construction versus 184.5 MW for renewables.[2] New Zealand needs to have switched the investment incentives from fossil fuels to renewable energy before the next potential wave of fossil fuel investment if carbon dioxide is to remain at the same levels in 2030. As discussed above and repeated here, if carbon emissions are to stay flat there has to be a 75 per cent penetration of renewable energy in 2030 – provided energy demand is at 1.5 per cent.

If renewable electricity accounted for 75 per cent by 2020, ten years earlier than in the previous example, and had reached 85 per cent by 2030, the proportion of fossil fuel to renewable energy capacity in the electricity system would have to fall from about 2012. With this higher proportion of renewables, carbon emissions start to fall and might reach the 3.3 $mtCO_2$ in 2030. This is equivalent to the emissions in 1990, which New Zealand has to get back to in order to meet the 2012 Kyoto commitment in 2030. It would have taken, under this scenario, 40 years to bring emissions back down to those levels. Europe and the UK are already talking about the necessity of reducing those levels by 60 per cent by 2050. New Zealand is beginning to significantly lag other parts of the world.

However, if a further wave of fossil fuel investment occurred in, or after, 2012, a 75 per cent renewable electricity target in 2030 would not maintain fossil fuel emissions at current levels and an 85 per cent target by 2030 would not reduce CO_2 emissions to 1990 levels. The final New Zealand Energy Strategy (NZES) published in October 2007 has announced a number of measures, including a target for renewables to provide 90 per cent of electricity supply by 2025. While not ambitious, it will require considerable change to the electricity mix. What are the underlying views and principles of the NZ Government which led to this policy announcement?

The competitiveness of wind energy in New Zealand

A key factor is an ideological rigidity about using pure market mechanisms. The Government's position, and *a fortiori* that of the current National Party opposition, is to encourage investment change through market mechanisms. The transitional measures paper published by the Government stated that 'although no decisions have been made, the Government has a positive view on the use of economically efficient price-based measures applied broadly across key sectors of the economy in the longer term (i.e. post 2012)' (MED, 2006), and this is widely expected to be a carbon trading scheme. Carbon trading is supported primarily because it is a 'least cost' mechanism and the 'quantity' to be reduced can be 'set' to where the Government wants it to be. A carbon tax has the benefit of being 'set' at the level which equals the cost the Government wishes to impose, but they cannot be so sure that it will hit the reduction target. Moreover, from a political perspective, in the 2005 election campaign, two minor parties which proved to hold a critical few seats in the House after the election, took against the carbon tax proposal, and it was dropped in December 2005.

In addition, there were two key sources of data concerning electricity costs for the Draft NZES, both of which are arguably very optimistic: the Energy Outlook 2006 (an MED report); and a report published by Meridian

Energy to coincide with the publication of the Draft NZES. These both showed that wind energy, new gas, coal and geothermal power plants had very similar NZ costs per kWh (NZc/kWh). If these figures were correct, carbon trading should switch the investment incentives from fossil fuels to renewables, thereby encouraging greater renewable energy deployment and thereby negating any need for any other mechanism than the pricing of carbon. Furthermore, two large energy companies announced that they would invest in renewables[3] thereby negating any particular need for any mechanisms in addition to the pricing of carbon.

However, these cost figures and the investment path they imply are arguably very optimistic. It's more that the *average* cost of combined cycle gas turbine (CCGT) generation appears to be cheaper than the *average* cost of wind generation (and other renewables). While the best wind sites are competitive with the average cost of gas, an average wind site will not be. In addition to this, wind energy suffers from a number of transaction costs or barriers to investment that do not apply to fossil fuel plants, thereby making wind energy a greater investment risk than gas. These barriers include:

- renewables are more capital intensive, so more upfront money per MW is at risk for any particular project;
- higher consent costs (under the Resource Management Act 1991) per MW;
- many projects don't get the full benefit of economies of scale because they are only 50 MW or so in size;
- there is a cost associated with the variability of renewables in relation to matching generation to a combined generator and retailer (known as a gen-tailer in New Zealand) retail portfolio from hour to hour.

Finally, it is unclear what the value of the risk is which arises from these barriers. To a large degree, the value of risk is unknown in New Zealand because so little wind generation has been developed, but in this can be expected to be high. Moreover, the value of risk will change depending on what mechanism is in place (i.e. carbon trading; renewable energy obligation versus feed-in tariff, and so on). However, the net effect of the risk and transaction costs is likely to be that a generator will only develop a renewable energy plant, rather than a gas plant, if it is confident that it will be paid a certain amount more per kWh, or receive an acceptable rate of return, than it would for the gas generation. The size of this risk and transaction cost will rise if the only mechanism in place is a carbon charge, and it will fall if there is a low-risk renewables specific mechanism such as a feed-in tariff.

Given the cost differential between gas and wind, and given the resource potential of wind at different costs, a value of carbon (derived from carbon trading) might raise the price of fossil fuels such that it switches the investment incentives from fossil fuels to renewables for certain high resource sites, but it is not certain that it will do so, and probably will not for lower wind speed sites. This means, at best, only a limited amount of wind energy from higher wind speed sites (around 10 m/s or more) will come forward and this will not be enough to halt the move from renewables to fossil generation or to increase the proportion of renewables in the electricity system, with the implication that carbon emissions will continue to rise.

NZ concerns

The rest of the chapter attempts to analyse the key issues which push against each other to deliver a rather ambiguous energy policy. In a nutshell, these are the NZ definition of a Good International Citizen (which one would imagine would lead to stronger environmental policies than are in place); the importance of 'cost' of the energy policy, but more specifically 'least cost' and the way that cost is measured; and NZ attitudes to innovation (which are conservative, linking innovation to economic development). Arguably, the second issue (that of the importance of cost) could also be filed as 'NZ attitude to innovation'. However, because of wider concerns of industry and the energy lobbies it is discussed as a separate issue.

An overview of the Government principles behind policies might, arguably, be said to be:

- that market mechanisms, while tempered and carefully looked at, are preferred and should be followed where possible;
- that quantitative economic analyses are central to Government decision-making and qualitative non-economic values have not been incorporated into mainstream Government thinking;
- that climate change is viewed as a technology, rather than an energy system, issue, so meeting the challenge of climate change does not necessitate a great deal of change or innovation – it is perceived as simply requiring 'new' technology;
- the view that technology can be 'bought in' at any time, so Government policy can be incremental and cautious to ensure no

mistakes and no undue cost, and reflects the underlying view that there is not really any urgency in reducing carbon emissions;

- there is little sense of climate change as an opportunity for New Zealand;
- climate change is not yet understood in New Zealand as a critical international policy issue and there is thus little sense that New Zealand's slowness in acting on climate change may tell against the country's 'brand' and ability to export;
- innovation theory concerned with systems and transitions has either generally not yet been taken up by academics, the exception being Jonathan Boston and Ralph Chapman of Victoria University (Chapman, 2004, 2006; Chapman and Boston, 2006), or has not yet made a mark on Government thinking.

Together, this leads to a conservative, cautious, incremental cost-based policy.

New Zealand as a Good International Citizen

New Zealand is very proud of its place in the world. It likes to think of itself as 'hitting above its weight' or some other phrase which implies that while it is a small country and only a fraction of the world's population, it is taken note of in a global setting because of the credibility it has built up over time from being a Good International Citizen. In addition, New Zealand is very proud of its 'clean and green' image. This is partly the result of its non-nuclear policies but also because, compared to most countries in the world, *it is* a gloriously unspoilt clean and green place. These two very important defining characteristics of New Zealand come together in its policy on sustainability and its position in the global debate on climate change. The Prime Minister, Helen Clark, gave a very pro-environment speech in Seattle in March 2007, which was widely seen to be giving a steer to the outcome of the NZES (Clark, 2007b).

The Draft NZES included a very impressive stakeholder engagement process. As part of this process, at a forum held to specifically discuss 'How far and how fast', and the value of targets, Jonathan Boston and Ralph Chapman put forward an argument that the definition of New Zealand as a Good International Citizen should be grounded in terms of planetary risk. They contended that New Zealand should have a policy which would 'do its bit' for meeting a parts per million (by volume) of CO_2 in the planet's atmospheric carrying capacity and that this global imperative should be the focus of NZ policy rather than NZ simply fulfilling its part of the international, politically set agreement of Kyoto. The argument was that merely sticking to Kyoto-type commitments would not be enough to earn the right to say that New Zealand is a Good International Citizen.

This is a very simple but clear idea which resonated throughout the rest of the stakeholder process. The recent pronouncements by the Prime Minister give the impression that NZ is much nearer the planetary than the international definition of a Good International Citizen. NZ official policy on the other hand looks towards the Kyoto-related definition.

The importance, and valuation, of cost

NZ industry has given, from the stakeholder involvement and submissions, very clear messages on its policy preferences for the NZES and NZ climate change policy. Any measures put in place should have had a detailed cost-benefit analysis undertaken on them to understand their cost-effectiveness, value and implications. 'Additionality' should be the centre of such an analysis, and if only limited additionality can be shown then caution should prevail and such a mechanism should be viewed as having limited value. Moreover, the calculation of that additionality should be by an economic analysis.

NZ industry would argue that if a mechanism has to be put in place then it should be 'least cost' and should mimic the market. New Zealand should only have to do as much as anyone else in the international process and no more. To do more, is to spend more and to possibly undermine the NZ economy more than is necessary. Furthermore, equity should be upheld between sectors. It is unfair, it is argued, that the energy sector has to reduce its emissions more than other sectors, for example agriculture.

Moreover, in general, industry exhorts the Government to think very hard about implementing any climate change policy at all since, it argues, climate change is a global problem and if the rest of the world does not fulfil its side of the agreement to the same degree as New Zealand, then NZ and its citizens will suffer unnecessarily. For example, so the argument goes, developing countries, including China and India, are going to be increasing their emissions (as they are allowed to under the Kyoto process). New Zealand is really a very small part of the global problem and is not a historically significant emitter and therefore why should NZ's competitiveness suffer? This view feeds into the other policies – the definition of good global citizen; the view that technologies can be 'bought in' when they are 'cheaper'; and the argument that the

Government should implement policies in an equitable manner across sectors so that the energy sector is not unduly penalized.

Finally, New Zealand sees itself as a trading nation, and industries and the Government are concerned to ensure that the competitiveness of NZ industries are not undermined by Government policies, including climate change and energy policies. Competitiveness at Risk (CAR) issues are very real given that many of New Zealand's trading partners are non-Annex 1 countries, which means that the latter will not have to bear the international price of carbon from 1 January 2008, when the Kyoto Commitment kicks in.

This is a brief summary of the position of the 'do-as-little-as-possible' lobby, comprising the fossil fuel-based energy companies; the major energy user groups and anyone else whose interest is to continue in the same paradigm. The lobby's importance should not be underestimated. Within the stakeholder process, it was a focussed, disciplined lobby which put out simple, clear and consistent messages. Moreover, and possibly more importantly, to a large degree the messages chimed with the views of much of MED and the Treasury.

This underlying view has a number of implications for the types of policies which can be discussed without attracting criticism from this lobby. For example, a clear view would be that the Government should not subsidize a renewable energy policy since many of the major energy companies had already stated that they would build renewable energy projects (Contact,[4] Meridian, Mighty River Power[5]). The industry lobby, and most other groups, would not support these companies being eligible for subsidies since they would argue that the Government would be paying for what a number of these companies (as state-owned entities) would do anyway: there would be no additionality from such a policy and NZ customers would have to pay more. This effectively rules out either a renewables obligation or a feed-in type mechanism.

However, the energy companies which have said they would develop renewable energy projects are only interested in developing very big power plants in very high (over 10 m/s wind energy) resource sites, since the generation from these locations is more or less competitive with new gas power plants. Only a few of the high resource sites for wind energy have reasonable access to transmission lines, and this considerably constrains the 'economic resource'. The large energy companies are not interested in developing the lower wind speed sites, because their returns would be too low (since there is no subsidy for renewables). If New Zealand wants to develop large quantities of renewables, as a means of increasing the proportion of renewables in the electricity supply mix and reducing the carbon emissions from that sector, it does have to find some means of ensuring that the lower resource sites are attractive to investors. The simplest way is to either subsidize renewable energy in general or wind energy specifically and accept in the short term that there will be limited additionality. A more complex policy would have to state clearly that only 'new entrants' were eligible for such money, something which would be legally difficult to police. Moreover, as the high wind speed sites are used up the additionality arguments alter. In many ways, it is sensible to get a renewable energy policy going to kick-start the currently very constrained industry. Certainly without a specific mechanism, renewable energy development will be limited and short-lived, at best. However, this outcome would not fit with the view of the industry lobby. If the Government wants to step outside and take on the industry lobby, it is going to have to be ready for a fight.

Another implication of the importance of cost, is that the NZ Government does not want to be tied in to a mechanism it might later want to get out of, if it is not successful or if it is too expensive. The Government is very risk averse. It finds it acceptable to become involved in carbon trading because the 'bet' is that it will become the 'global long term' means of trading carbon. Other costs, such as those which derive from a specific mechanism (e.g. for renewable electricity), are less acceptable partly because of cost but also because it may not be successful.

Another key aspect of the cost approach adopted is that it is very static. Costs are not seen as responding to learning by doing. Little recognition exists of the economic literature around innovation and transition paths based on learning and cost-reduction (e.g. Arthur, Rotmans). And little acknowledgement is given to voices arguing for experimenting with a deliberate transition that aims to open new learning opportunities, and reduce costs, over time.

Incremental and cautious least-cost approach

New Zealand has to date adopted an incremental approach to energy policy. It is likely to put in place a carbon trading scheme, which will alter the price of carbon and competitiveness of fossil fuels and renewables. The Government will then wait and see how that works out and if enough renewables are perceived to be coming through to stabilize carbon emissions, it will feel vindicated in its policy caution. If not enough are coming through, the Government will put in place another policy which will then address that particular problem. It can then argue that it has not wasted money and it has implemented policy from evidence in a rational manner. Thus, no undue cost occurs; only additionality happens;

no company receives money the Government could otherwise have spent on something else. All this appears very sensible, except if an important goal is to reduce CO_2 by a certain amount by a certain time (i.e. there is urgency), and except if another goal is to encourage a more dynamic and innovative energy sector. In other words, if there is more urgency to the climate change challenge than the cautious approach demands, and/or there are other economic development challenges to which climate and energy policy should/might have regard.

View that it is a technology issue not a system issue

Lack of urgency because of the ability to 'buy in' a technology

An important attitude in New Zealand is that they are 'technology takers'; they are not technology innovators. This is part and parcel of the cost argument which runs: New Zealand is a small country; it cannot input resources into technology development to the same extent as the bigger, global economies can; so it is better to wait and see which technologies succeed in the global world rather than 'picking technological winners' themselves; once it is clear which technologies have 'won', they can be 'bought in'. It is a least-cost approach. New Zealand can wait and see and not spend money. This of course relies on other countries paying the development costs and taking the risks. It also means that New Zealand does not benefit from these new technologies, and this is discussed below.

From a country perspective, the view that technologies can be 'bought in' when other countries have developed and demonstrated them, represents a 'technology' view of innovation. Somehow a technology can be incorporated into an energy and electricity system easily, requiring little change to the wider energy or electricity system.

An example of this is the policy towards carbon capture and storage (CCS). Clearly, a viable CCS technology would be hugely beneficial to New Zealand which has a small amount of electricity capacity which is dominated by one large, modern, coal power plant, Huntley. If CCS could be attached to Huntley, New Zealand's future carbon emission trajectory would be hugely improved. However, the economics of CCS are currently poor, except in very one-off situations, such as Sleipner in the Netherlands. Many countries are investing in CCS technology, for example both the UK and China have either announced or are building demonstration plants. New Zealand is therefore hoping that it will be able to 'buy in' the technology at some later date. Given that possibility, one argument could be that a long-term renewable energy target, for example 85 per cent by 2030 (or like those advocated by the European Commission or Jonathan Boston and Ralph Chapman) has the potential to undermine the development of CCS in New Zealand. In one sense, this cannot be argued against since, if there were an 85 per cent target for renewables in 2030 it would mean that CCS would have to be economic with the remaining non-renewable 15 per cent of the electricity system. This is clearly going to be harder than if it was trying to be economic within the current 30 per cent fossil generation.

However, who knows what demand will be in 2030 or what the cost of other technologies will be or indeed if 'other' unknown new technologies will have come forward. A logical, low-risk policy for New Zealand is to support both renewables and CCS development, to the extent it can. CCS is not expected to be technically viable before 2025, and it could be much longer off, if at all. New Zealand requires low carbon alternatives now and has a brilliant renewable energy resource with technologies which are nearly competitive and mature, *now*. If both renewables and CCS succeed, the country will have increased its low carbon options; and if only one succeeds, then the other is there as a back stop. However, if New Zealand does not seriously try to develop renewables it is essentially relying on (picking) a far more immature, risky technology than the relatively mature wind energy that is available now. The question is whether the overall benefit to energy policy by establishing a 2020 target for renewables outweighs the dis-benefits to CCS, because of its reduced market for CCS technology.

Nevertheless, it seems that New Zealand prefers the 'wait and see' approach to what technologies work. In conjunction with the incremental approach, the Government can argue that it doesn't necessarily need to support new technologies if carbon trading shifts the investment incentives. A benefit of CCS is that it suits the underlying characteristics of the present energy system. 'Buying in' a technology reflects a fundamental fault-line in attitudes to innovation. If a country believes that it is better to wait until another country develops a technology it is essentially saying a number of things.

It is saying that it believes that it can bring in a technology, wholesale. This might be the case to a lesser degree for certain technologies but energy technologies have to work within a system. That system has environmental standards, economic regulations, electricity system security standards, legal requirements, planning permission procedures, human skills required to run the energy plants, and all the various other requirements to build and keep the energy plants going. The development of this framework which matches each different energy technology takes time to develop. Even when effort is made to make changes, it takes time for electricity system security standards to go through the processes of change; for health and safety rules to be re-written and agreed for different technologies and then to be incorporated with the previous rules; for environmental regulations concerning the plant wastes or emissions to be developed, and so on. 'Buying in' reflects a linear, rational attitude to energy system development but one which doesn't reflect the real world.

It also means that money doesn't have to be spent now. That means no money has to be spent on R&D, universities, student courses, etc. This has a wider effect in that New Zealand has very few skills in this area; it also has very few people arguing for different energy pathways to Government as a result. Government money does not have to be spent now, so the industry lobbies are reasonably supportive of Government; the Treasury is happy, the energy system can continue as it always has done so those that work in it are reasonably happy. This fits with the cautious, incremental, non-urgent approach.

What it does not fit with is the definition of Good International Citizen. Nor does it fit with the case for urgency. Thus, the current NZ energy policy is also saying that climate change, and their Good International Citizen 'bit' towards, it is not urgent. Only when the Government considers climate change as an urgent global issue will it have to confront the underlying and comfortable principles and policies by which it exists.

Does not believe in value of industrial policy (too close to picking a winner)

Another parallel strand of NZ energy policy and being a 'technology taker' is that New Zealand does not have a supportive industrial policy in relation to energy policy, although arguably it does for agriculture. It sees itself as an early adopter but not a developer – and partly this is understandable given the amounts of money other countries are investing in new technologies and because of its views of being able to 'buy in' technologies as and when it pleases. Unlike any European country, New Zealand does not appear to view its wonderful renewable energy resource as a resource. More, it feels comfortable with the position it is in – i.e. that 70 per cent of electricity is supplied from hydro – and therefore considers itself already ahead of the pack. The fact that NZ emissions are at twice the 1990 Kyoto level with high annual rates of energy demand does not seem to have fundamentally dented that view.

New Zealand has not used the review of its climate change, energy and land use policies to link up different sectors or the macro-economic needs of the country. For example, the agriculture sector provides around 50 per cent of the climate change emissions, mainly from methane. It would therefore seem to make a lot of sense to try and restructure agriculture somewhat from food to energy. For example, this might be biofuels, forestry or farming the wind. There are a great many sensitive issues when dealing with this sector as a result of the enormous consequences wrought on it by removing agricultural subsidies in the late 1970s and early 1980s. Even so, there are opportunities which have not as yet been taken. Moreover, New Zealand has one of the best marine resources in the world and could become the southern hemisphere centre.

Conclusion – What does all this boil down to for the importance of political paradigms?

The intention of this case study was to illuminate how all countries have a political paradigm and that they act as a shaper, a constrainer, and a channeller of policy development. New Zealand has a very particular character and is very independent but its underlying political paradigm is similar to the UK. It is however a country of the southern hemisphere and South Pacific, a long way from Europe and its policies. It is conducting its policies according to its neighbours and needs, and the end result is an energy policy which is very limited compared to those being put in place in Europe. At the moment, the paradigm is constraining policy to carbon trading. It will be interesting to see what happens when the sense of urgency about climate change reaches New Zealand, because it does have a very real sense of its 'clean and green' credentials and it may be that the popular demand for more intervention will be enough to push it in to a more supportive policy. Unlike Britain, it does not have the powerful, interwoven web or 'band of iron' keeping the regulatory state in place. In this sense, it should be much easier for New Zealand to move forward.

8
Examining European Political Paradigms

The previous three chapters have explored in depth the sustainable energy policies and economic regulation in place in the UK and New Zealand. The intention has been to show that the overarching political paradigm is fundamental to the type of policies (direct from Government via Parliament) or markets and regulations which emanate from institutions which are set up according to those principles and which are representative of the paradigm.

Making the transition from a 'dirty' to a 'clean' system requires change across the energy system, whether it be the policies, the economic regulation, planning laws, banking and finance rules, or consumption behaviour. The success of those policies and regulations in delivering firstly the narrower (and easier) sustainable energy policy as part of the wider (and more difficult) sustainable energy system will depend on the extent to which a country is able to build up a connected framework which provides the support required for such a complex transition. This chapter expands on this by discussing the various sustainable energy mechanisms at play in Europe and asking why it is that some countries 'just do it' and put in place policies which reduce hurdles or raise technology development above the barriers (e.g. Germany and Spain). Other countries have had this 'just do it' policy in the past (e.g. Denmark) but have altered them in order to link them (or bring them more in line) with market requirements and, to a large extent, are seeing their renewable energy deployment plummet. Other countries (e.g. the Netherlands) have never quite managed a 'just do it' attitude and struggle, as the UK does, to combine concerns for the environment with a strong pro-market ethos.

The most successful policy in terms of renewable energy delivery is the feed-in tariff (FIT). At root, it guarantees prices per kWh to investors, thereby giving confidence and the ability to obtain finance unlike the unpredictable ROC prices. The result is lower investment costs:

> Contrary to criticisms of the feed-in tariff, analysis suggests that competition is greater than in the UK Renewable Obligation Certificate scheme. These benefits are logical as the technologies are already prone to considerable price uncertainties and the price uncertainty of tradable deployment support mechanisms amplifies this uncertainty. Uncertainty discourages investment and increases the cost of capital as the risks associated with the uncertain rewards require greater rewards. (Stern, 2006)

It is widely seen as a better approach than tradable quotas like the Renewables Obligation, although of course its success means that it leads to more capacity and higher overall charges to consumers:

> Comparisons between deployment support through tradable quotas and feed-in tariff price support suggest that feed-in mechanisms achieve larger deployment at lower costs [although] greater deployment increases the total cost in terms of the premium paid by consumers. (Stern, 2006)

In addition, in some schemes the prices paid are degressed (i.e. adjusted downward) to match the technology learning curve, thereby avoiding the high prices as compared to the RO which gives out payment that is not needed and is, in effect, profit. Overall then, the FIT approach thereby promotes dynamic efficiency (Mitchell et al., 2006).

However, generalizing about different countries is fraught with difficulties. Different countries have undertaken similar policies but which have produced different outcomes (Dinica, 2006). Every country is complex and it is all too easy to pick out one 'fact' in support of an argument while leaving out other more salient 'facts', either through design or ignorance. The discussion in this chapter is purposefully high level, or general, with the intention of drawing out which side of the innovation fault-line the various countries lie on. It does not try to explain in detail each country's policies but to highlight what each political paradigm has had to find acceptable in order to establish each policy, or set of policies.

Innovation and sustainability

Chapter 2 developed a key idea of the book – that there is an innovation 'fault-line' and countries which are on one side of it are likely to be better at delivering sustainable energy policies and a sustainable energy system than if they are on the 'wrong' side of it. The innovation fault-line is:

- an understanding of what 'innovation' is and that not all of it is 'good';
- an acceptance that markets are not the best way forward for making *all* choices – although certainly they are for many decisions (if not the majority) and will continue to be central to any future sustainable energy system;
- that it is not only acceptable to 'pick' a technology to support but that it is necessary to 'channel' innovation policies towards sustainability;
- that choosing to support an environmental option, which may not be a least-cost measure, rather than choosing the economic or market option, may be appropriate, necessary and sensible and provide a great deal of additional value, albeit not in a way which can readily be valued monetarily;
- accepting that trying to meet the challenges of climate change is a 'system' issue not a technological-only issue.

An important aspect of this is that not only do these attitudes to innovation matter but they all have to be present. It isn't a 'pick and mix' situation. If a country is on the right side of the innovation fault-line and is trying to connect all aspects of its innovation policy, the details are less important. It is not the case the other way around.

The notion of a transition to a sustainable energy economy, and the ability to make it happen, is a highly contested area (Shove and Walker, 2007; Smith, 2006) as was discussed in Chapter 3. This book is about applied policy making, although it has tried hard to link practice and theory. Given the academic disputes concerning the ability to stimulate innovation or undertake a transition, and with apologies to those who might consider the following discussion too simplistic:

- the transition to a sustainable energy system *is* a system issue and not only about reducing carbon and other greenhouse gases (through reducing energy demand or increasing (energy) efficiency

of resource use). It additionally requires, inter alia, the development of (new) low carbon supply technologies, new institutions, new behaviour and consumption attitudes, and probably unknown 'new' factors;
- innovation is not predictable or linear. Some policies are more predictable than others, to the extent that an outcome can be incentivized (for example, carbon reduction via carbon trading) even if the indirect effects are as unknown or as unexpected from other policies. However, least-cost and incentivized policies, of which carbon trading is one, tend to be prescriptive which limits innovation. This implies there are two sorts of policies – carbon reducing policies (which can be incentivized and least-cost, such as carbon trading) and the wider low carbon energy system development policies (which can be least-cost but are more likely to be successful if they are not).

This implies that stimulating innovation across the energy system is related to:

- encouraging an environment that is conducive to 'opening-up' possibilities, and this includes putting in place policies and economic regulations where the absolute outcomes are not 'known' (for example, the total cost per year is unknown; or the total number of projects to be delivered is unknown). This enables a space for innovation, unpredictability, and/or unexpected side-effects to occur; and in tandem with this
- implementing policies or economic regulations intended to reduce hurdles by ensuring the incentive is open to all without competition. This means that companies which are larger and have access to economies of scale, or which have been around longer, are unable to exclude individuals or smaller companies immediately. The individuals or smaller companies are able to participate, with unknown outcomes.

An introduction to the feed-in tariff (FIT)

As discussed above, feed-in tariffs are a very effective way of deploying renewable energy. Feed-in laws, which provide a feed-in tariff or payment, are one of the two main mechanisms used to promote renewable energy around the globe. They are increasingly popular – for example, 18 of the EU's 25 member countries use FITs. The basic format is that an obligation is placed on electricity grid operators to buy the renewable electricity at a pre-set price and to give it priority access to the grid. A recent spate of articles, including a European Commission Review, have confirmed that feed-ins have been more effective in promoting renewable energy than obligations and more cost-effective (EC, 2005a; Carbon Trust, 2006; Szarka and Bluhdorn, 2006; Elliott, 2007a). This is because, unlike quota/obligation schemes such as the UK Renewables Obligation, in which prices are unpredictable, the guaranteed/premium prices reduce investor risk, and the process of price degression over time adopted in many schemes avoids overpayment as the technology matures. They are also accessible to any investor, and are technology specific, which leads to a diversity of technologies, investors, new entrants and geographical resources, thereby enabling support for both large-scale marine technologies as well as micro-generation. And they are also administratively simpler.

The two countries which have been most successful in delivering new renewable energy capacity, Spain and Germany (see Figure 8.1), both have feed-in tariffs and both are discussed below. However, it is also possible for feed-in tariffs to be unsuccessful if they are badly designed.

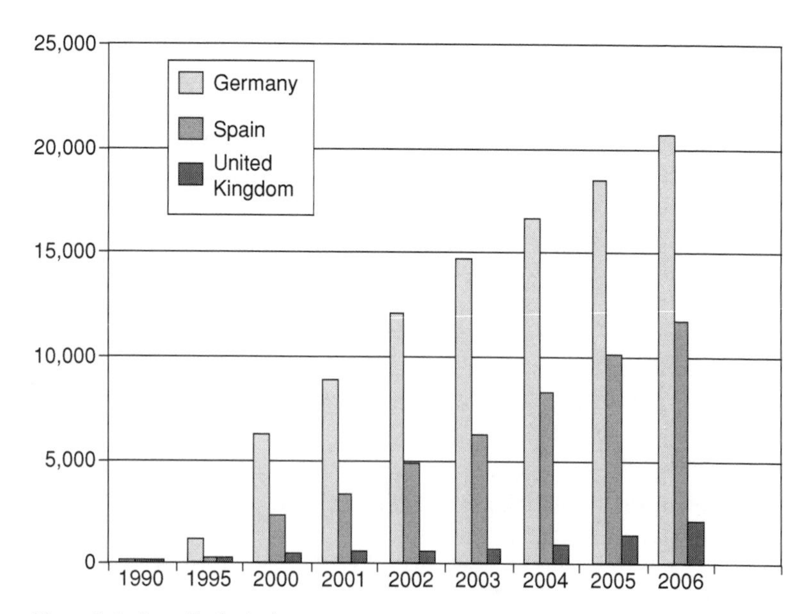

Figure 8.1 Installed wind energy capacity in megawatts

Source: WindStats

Successful feed-in mechanisms have very low levels of risk. The key to a successful policy is to minimize risk and provide certainty in each section of the policy:

- a high enough payment to make an adequate return;
- the way the payment (and its revisions) is determined must be transparent;
- the electricity purchase obligation must guarantee priority access, be simple to put in place, and be of long enough duration;
- the rules of connection to the grid must be simple;
- paying differing prices for different technologies enables diversity;
- ensure the equal burden sharing, or distribution, of the costs of the feed-in law across all electricity consumers, by including the costs in the power price so that no particular area pays a higher cost than another area; and
- establish parallel long-term targets for renewable energy to bolster investor confidence.

When successful, feed-in laws have a number of beneficial properties:

- they can be very successful in terms of installation of capacity;
- the capital costs of investors are minimized due to the low level of risk;
- there are low administration and transaction costs;
- no market liquidity problems;
- their stability helps to underpin high-quality components and, in the case of Germany and Spain, world-class export industries;
- they promote a diversity of technologies (from large, immature technologies, such as marine technologies through to small, mature micro-generation technologies);
- they enable a diversity of investors (very large companies through to individuals) thereby enabling new entrants;
- a geographical diversity (meaning dispersed and different resources) which benefits a spectrum of localities; which complements the least-cost approach to network development; and which reduces conflicts concerning planning permissions, even if it does not overcome them entirely.

Details can be tailored to individual countries. For example, Greece and Portugal include a 2–2.5 per cent payment from the generators to the local municipality, as a means of increasing local support; and in Slovenia the payment for electricity is higher for certain peak times of the day.

The cost of the measures is paid for through electricity bills. Feed-in laws currently apply only to electricity, but could also be applied to heat. The cost of the mechanism is directly related to the amount of generation bought and the level of the payment paid. Germany is equal first with China in terms of total investment in renewable energy and has a clear policy goal of developing international renewable energy industries, as discussed below.

A means of by-passing barriers

From the perspective of Governments which want to 'know' the cost of every Government action, FITs present a difficulty. A 'pure' feed-in mechanism, whereby a price is 'posted', means that the Governments cannot 'know' what the maximum cost in a year will be to them because they do not know how much capacity will be installed, particularly in the first year. The second year onwards can gain insights from the previous years.

A cap can be placed on the FIT, for example 1,000 photovoltaic roofs. This was the way Germany started out with its policies and, in this way, the annual maximum total cost can be known. An early Dutch mechanism was similar. However, in this case it did not work well because the applications for the year's capacity was used up on Day 1 of the year, and led to individuals being paid by companies to camp outside the relevant office for weeks at a time to ensure that they were first in the door on the application.

Capping a FIT, while possible, represents a fundamental constraint on what a non-capped FIT represents in the way a country supports sustainability. If a well-designed, non-capped feed-in is established (as in Germany and Spain), the country is saying: we want renewable energy, we want sustainability, and we'll pay what we must to get it. This attitude – or determination – has indirect (unquantifiable) effects of providing confidence to the investor community. This sign of 'political will' is vital for providing investor confidence, but not something which has ever been captured in the UK.

Feed-in laws are also extremely flexible. It is possible to change payments for technologies; change lengths of contracts for new contracts, because there is no link between the past and future. This means that technology developments or new requirements can be incorporated easily. This is very different from obligations which are inflexible, because of the complex valuation of ROCs extending into the future.

One false criticism of the feed-in laws is that payment does not provide an incentive for using the best resources first or to be sited in appropriate places for the grid. Payment is for generation and therefore the use of the best resource sites will maximize payments. Thus, in economic terms the incentive would still be to go for the best resource sites. However, the feed-in simply enables a wider resource base to be used than competitive obligations (as explained below), and this is beneficial for two reasons. Firstly, good resource sites are often in beautiful areas or sites of special cultural significance and their use may cause opposition, so the ability to go elsewhere is helpful and makes it easier, and less contentious, to obtain planning permission. And secondly, several studies have shown that the cheapest way to incorporate renewables in to an electricity system is for them to be dispersed across the system in terms of scale, technology and geography.

Another criticism of feed-in laws, is that they are not (as) complementary to electricity markets as obligations. As mentioned above, and discussed further below, Spain and Germany have been the most successful examples of feed-in laws. They represent the two different types of design: Germany is based on a fixed, pre-known payment which differs for different technologies. Spain's mechanism offers two modalities for renewable electricity generators to choose from: a fixed tariff (as occurs in Germany) or a market price plus a premium payment plus an incentive to participate in the market. Generators can choose annually which modality they prefer. The premium payments are revised every four years. The incentive to participate in the market price is 10 per cent of a reference electricity market tariff, which is calculated annually. Given current prices for electricity, most producers choose the market + premium option, but this depends on the perceived risk of the choice in any given year. This means that the renewable energy generators do have to take note of the electricity market to work out how to maximize their payments. In neither case, however, do they interact in the electricity market because of the guaranteed payment (buying), or priority access, of their electricity by the electricity company.

This priority access, and concomitant reduction in risk and accessibility by any potential generator (of any size and type), is fundamental to the feed-in law. Moreover, both Spain and Germany have rules of connection to the grid, and its cost. This by-passes, but does not do away with, the development of appropriate network access rules and incentives. FITs both encourage investment but also encourage investment from new sources of capital (or new investors) because it makes it easy to connect and sell the electricity. This reduction in risk makes a feed-in tariff what it is. It is the opposite of the short-term goals of renewable energy development. The UK wanted to force the renewable energy generators to form a greater part of the electricity system and did not like that renewable energy generators were, in some sense, separate from it. However, the FIT doesn't do this, accepting that this will happen once the ten-year FIT comes to an end.

However, there are also wider benefits of FITs which the obligation method of supporting renewables lacks (Mitchell et al., 2006; Elliott, 2007a). There are costs to the electricity system of integrating renewable electricity. The cheapest way of doing this is, as described above, by dispersing them around the system in terms of size and technology and then to have one 'balancer' of the system. This occurs with a feed-in tariff but not with obligations. Similarly, its non-confrontational aspects to planning permission are helpful.

The 'just do it' countries

Germany and Spain are both countries which have had phenomenal success in deploying renewable energy (Ragwitz and Huber, 2006). Germany is the well-known architect of the 'classic' FIT. This has become increasingly sophisticated since its original law in 1991. In addition to the FIT payments for different renewable energy technologies, there is also the ability to borrow money at cheap rates (for a variety of uses, including renewable energy development) combined with tax exemption for investment in renewable energy projects, which is particularly attractive to higher earners. FITs promote diversity of investors, technologies and resources. By their nature, they require that grid operators connect the renewable energy projects according to clear rules, and this has altered the skills and attitudes of the grid operators to renewable energy. However, not only was support for renewable energy in Germany to alter the basis of electricity generation but was also seen as a new direction for industrial policy.

As Szarka (2006) explained, the German policy was based on an interconnectivity:

- in the early R&D phase, policy encouraged technological variety;
- in the later phase, it encouraged market creation and development by bringing many investors to the market;
- an industrial policy fostered a domestic equipment industry;
- the policy built on and encouraged the social legitimacy of wind.

However, Germany has also reflected inter-connectivity across the energy system and in this sense is a country on the right side of the innovation fault-line. The large German energy companies have consistently complained about the FIT and took the Government to court about FIT requirements, until 1998 (Stenzel and Frenzel, 2007). The large energy companies have an extraordinarily low ownership of the wind energy. This was originally because utilities could not own more than 25 per cent of a project and were not able to develop projects in their own monopolies. This meant that they did not build up their own skills in relation to renewables. Moreover, energy demand was more or less flat in Germany from 1990 to 2002 (IEA statistics), so that any new generation was displacing in-situ generation, owned by the large companies. The restrictions on ownership were removed in 2000.

Despite the utilities' opposition from 1990 to 1998, a Social Democratic/Green Government was voted in and solidified support. It is an example of a country which is prepared to intervene – whether it be in support of particular technologies; to ensure grid connection; and within the market to ensure priority access. The FIT started off in 1990 relatively cautiously with a cap both on photovoltaic roof and wind energy. It has developed in sophistication over time to incorporate such features as sunset clauses, payment reductions in line with technological gains, and so on. The German Government has been prepared to set a price and then pay, via consumers, for however much that premium payment was taken up, without knowing beforehand how much this would be. The Government, took the risk. While the payment per kWh is lower for wind energy than in the UK, because the FIT has been so successful in terms of encouraging investment, the total cost is higher than in the UK.

In 2006, 73.9 billion kWh (TWh) or 10.2 per cent of electricity generation in Germany came from renewables, of which two-thirds was covered by the German Renewable Energy Sources (RES) Act 2004 (which followed on from the Renewable Energy Sources Act, 2000 and the 1991 Electricity Feed-in law). Thus was a reduction of 68 million tonnes of carbon dioxide. The fee payments to plant operators (generators) totalled €4.1 billion. The resulting additional cost (the so-called differential cost) as compared with the costs for electricity generated from conventional energy forms equalled around €2.4 billion. This equates to 3 per cent of the price of electricity paid by domestic customers. The differential cost triggered investments of approximately €5 billion and there are now around 240,000 people working in the renewables sector in Germany (BMU, 2007).

Spain has also been successful in developing renewables, albeit in a very different way from Germany. Spain has provided a dual modality for investors. They can either choose a standard FIT payment or a payment linked to the electricity market. The usual requirements of the FIT apply: such as priority access; rules for connection and payment to the grid. Unlike Germany, the Spanish utilities were always allowed to invest freely in renewable energy and benefit from the FIT. Energy demand was also rising so that renewable energy generation was not displacing their own generation. This meant that the energy utilities were able to develop their own skills of network connection. As with Germany, the renewable energy programme in Spain has led to several companies that have become very successful both domestically and internationally. In Spain, Gamesa has become the dominant wind turbine developer and the second most successful global wind turbine developer (behind Vestas), on the back of its domestic programme. Gamesa is a subsidiary of Iberdrola, the second largest utility in Spain. Thus, unlike Germany, renewables have become integrated and valued within the Spanish mainstream electricity system. This has had benefits. For example, Iberdrola has collected the data from their wind turbines to better understand how they should design and manage their system.

Table 8.1 Diffusion levels of wind power generation in the United Kingdom, Germany and Spain by the end of 2005

	United Kingdom	Germany	Spain
Total installed wind capacity	1,353 MW	18,428 MW	10,028 MW
Share of wind power in total electricity generation	0.45%	4.3%	7.78%
Total installed watts per capita	15	209	202
Largest capacity of big utility	RWE npower 395 MW	E.ON 224 MW	Iberdrola 3,494 MW
Share of all utilities in wind energy	81.89%	1.21%	58.35%

Source: Stenzel and Frenzel (2007).

As Stenzel and Frenzel (2007) show, the developers in Spain are very different from those in Germany and the UK. The incumbents see their future clearly linked to renewables and have taken it up with enthusiasm. Spain has managed, in a way that no other country has, to encourage investment across society. This has been achieved partly through FITs, partly through industrial policy but also through a much more positive attitude to renewables from the energy utilities because they could benefit from the FITs. Again, Spain has been on the right side of the innovation fault-line. It has been prepared to intervene for specific technologies; with respect to grid connection and markets; and for regional and domestic policies.

The UK system which channels its capped, risky mechanism through incumbents, thereby excluding new entrants, appears to be the worst of all worlds. It is the combination of the cap (which would if it was in any way reached stop development), the low buy-out price, the risk, and the exclusion of potential competitors, which enables the UK energy companies to develop at the speed that they wish. In Germany, the low-risk FIT has attracted new entrants and been successful, but it has not involved the energy utilities. In Spain, however, it has involved the energy utilities while at the same time enabling new entrants.

Making the transition to a sustainable energy system requires a series of policies across the energy spectrum to, inter alia, stimulate technology development, consumption behaviour, and deliver appropriate institutional arrangements, such as economic regulation. This section has provided a very brief overview of Spain and Germany. They have both managed to develop powerful renewable energy industries in tandem with having energy agencies and economic energy regulators. How this has occurred, unlike in the UK and the Netherlands, is discussed below.

Denmark as an example of a country which has lost the ability to deliver

Denmark was an early pioneer of sustainable energy policies, along with America, in response to the 1973 oil crisis. It has followed a very individual policy (Meyer and Koefoed, 2003). Not only did it promote sustainable energy policies before most other countries, but the mechanisms it used were different and closely related to the national characteristics of Denmark. Risoe, an energy agency, published data of wind technology performance from any wind turbines which had in any way benefited from the Government's renewable energy stimulation programme in the early 1980s. Since payment was related to the output of the wind turbines, the stimulation programme led to the development of sturdy and reliable wind turbines, which proved to be the basis of future international export success. The years up until the early 2000s provided payments to renewable energy which, while not a FIT, were a reasonably risk-free payment derived from a combination of an energy value and a payment for its carbon-free nature. While the energy payment tracked electricity prices, and so the final payment was not absolutely certain, it did provide enough certainty to obtain finance and stimulate investment. The rules for connection to the grid, and payment for it, were clear and easy to follow. Moreover, the mechanism enabled Danish householders to invest, with tax-free returns. Together, this led to the development of a powerful body of primarily domestic mentors for renewable energy in Denmark. This drove significant capacity installation and, in turn, underwrote the Danish dominance of global wind energy markets (Buen, 2006).

Against this background, the political leadership changed in 2000, as did the mechanisms in support of renewables. From the relatively risk-free payment, Denmark moved to a tradable obligation and this has led to a substantial drop in domestic renewable energy installation. The year 2001 is notorious within Danish wind energy history as being the year when almost no wind energy capacity was added at all. Although the Government is saying it wishes to turn this around (Transport- og Energiministeriet, 2007), as yet, there is no sign that this has happened. Denmark was a classic example of a country on the right side of the innovation fault-line, which set about implementing sensible policies which led to renewable energy delivery. It reduced risk by establishing a clear payment; it 'opened up' the environment both by the clear payment but also by enabling measures for domestic households; it made it easier for households or communities to become involved by publishing data about the different technologies; it simplified the connection of renewables to the grid; and it expanded a mentoring class of the public in support of sustainable energy in particular and sustainable development in general. Together, investing in and deploying renewable energy became a straightforward experience. However, that ease has disappeared as have the high rates of wind installation.

A different policy – the Netherlands and transition theory

The Netherlands has had a very different policy towards sustainable energy, from the rest of Europe, since the implementation of its early policies in 1990 (Agterbosch et al., 2007; Agterbosch et al., 2004; Aubert, 2007). Like Britain, it has a fundamentally pro-market political paradigm which is uncomfortable with market intervention and technology-specific support. While it has operated a FIT since 2003 (van Rooijen and van Wees, 2006), renewable energy development is still limited. Van Rooijen and van Wees are very critical of the Ministry of Economic Affairs' poor appetite for taking advice from stakeholders, arguing that had it done so the Netherlands might well have done better in delivering capacity.

One of the most interesting aspects of the Dutch sustainable energy policy, is that it has endeavoured to integrate transition management into Government policy (Kemp and Loorbach, 2005). In 2001, the Dutch fourth national environmental policy plan (NMP4) argued that there are seven barriers to sustainability:

- Unequal distribution – poverty causing irresponsible environmental management
- Short-term thinking (in politics and business)
- Fragmented policies and institutional deficits
- Prices which do not reflect external costs of environmental degredation
- Actors causing problems which do not own the problem (i.e. they are not responsible for the solutions)
- Insufficient precaution
- Solutions involving systems are surrounded with great uncertainty.

System innovation, meaning the ability to make fundamental change in functional systems and product chains, became the focus of policy. The Netherlands, along with British (SPRU and Imperial) and Scandinavian (Chalmers) academics, has the biggest concentration of transition management thinkers. The Dutch academics managed to permeate into the energy policy world so that Dutch energy policy has incorporated transition management (Kemp and Loorbach, 2005). This is both unique and welcome in the sense that it is innovative in itself despite questions of its ultimate value (Kern and Smith, 2007; Shove and Walker, 2007).

From the perspective of political paradigms (or political landscapes), the Dutch Government has, like the UK's, struggled to unequivocally support sustainable energy policies. Like Britain, it has endeavoured to do so within the confines of the market. The Netherlands, along with Britain, has been at the forefront of examining ideas for appropriate network configurations and payments, via its economic regulation, to encourage sustainable energy (Sustelnet, 2005). It seems as if, in both the UK and the Netherlands, in the absence of a supportive political paradigm, that the sustainable energy industry or academics in support of sustainability have tried to find ways of arguing for sustainable outcomes within the confines of the market approach. More efficient connections and economic regulation suits both sustainability and efficient economic regulation, as does the transition management type arguments of innovation policy. From the perspective of this book, transition management theory is welcome because of its overt acknowledgement of the importance of the energy system. Understanding how all the different parts of the energy system interact with each other should be valuable when operationalizing the transition from one system to another. As described in the final chapter, at root, this book believes that a powerful, interwoven framework of policies has to be put in place to move from the current situation to a sustainable energy system. This requires political decisions and actions. It seems to this author that while transition management ultimately cannot undertake those actions, it may be important in deciding to take those actions, and if this is the case, it is valuable.

Even so, the Netherlands has taken a more pro-active stance to innovation than the UK. It does now have a set payment per kWh: a FIT. However, it does not include priority access or clear rules on connection and payments to the grid. In this way, it is not so risk-free as the German or Spanish versions of the FIT. As with Germany and Spain, the Netherlands has both an energy regulator and an energy agency and their importance is discussed below.

The Netherlands also has a General Energy Council, as opposed to its Energy Agency (Novem), whose mission is

> an advisory board for the Government and Parliament of the Netherlands on matters of energy policy. The Energy Council aims to serve as a conscience for Government and society to contribute to the public energy debate, with the public interest always as its central concern.

This idea of having a 'conscience' for Government and Parliament on behalf of the public interest is also unique and of interest.

Harmonization of EU renewable energy policies

The European Commission would like to harmonize the various support structures as part of a process of integrating energy markets EU-wide. FIT schemes are seen as a block to this, since as set up they do not have a tradable function. The EU, on principle, prefers the tradable obligation schemes, such as the RO since this should lead to least-cost development of renewable energy capacity across Europe. In January 2005, the EC introduced an Emission Trading System (EU-ETS), a 'cap and trade' arrangement similar to the RO, but based on a carbon cap and trading carbon credits, rather than energy quotas and tradable renewable energy certificates (ROCs) as in the RO. In time, the ETS is seen by the EU as being one of the main mechanisms for driving renewables forward because, as the caps tighten, the incentives for low carbon generation should increase.

In the long term, existing renewables may not need FIT-type subsidy systems, and should be able to operate in a competitive market environment which reflects the value of the emissions they help avoid, of the sort created by the EU-ETS. For the moment however, the EU has had to accept that the various national FIT-type systems will continue in parallel because they have been so successful. For example, in 2005 in its second progress report on the EU Renewables Directive, the European Commission admitted that feed-in tariffs were 'currently in general cheaper and more effective than so-called quota systems, especially in the case of wind energy', and it accepted that it would be 'premature' to attempt to impose a single 'harmonized' renewable energy support scheme at this point. It even claimed that having a variety of national schemes 'can be healthy in a transitional period, as more experience needs to be gained', although this was still seen as an interim stage, before a single EU-wide scheme could be selected (EC, 2005a).

Clearly the EC still saw harmonization as the long-term aim: 'The integration of renewable energies in the internal market with one basic set of rules could create economies of scale needed for a flourishing and more competitive renewable electricity industry.' And it saw REFIT schemes as problematic. Although it said: 'These schemes have the advantages of investment security, the possibility of fine tuning and the promotion of mid- and long-term technologies', it argued that 'on the other hand, they are difficult to harmonise at EU level, may be challenged under internal market principles and involve a risk of over-funding, if the learning-curve for each RES-E technology is not build in as a form of degression over time' (EC, 2006). However it admitted that

> harmonisation through a green certificate scheme with no differentiation by technology would negatively influence dynamic efficiency. Since such a scheme would promote cost-efficiency first, only the currently most competitive technologies would expand. While such an outcome would be beneficial in the short run, investment in other promising technologies might not be sufficiently stimulated through the green certificate scheme. Other policies would thus need to complement such a scheme.

Moreover, 'a European wide common feed-in scheme which takes into account the availability of local resources could drive down the costs of all RES technologies in the different Member States' (EC, 2006).

To outsiders or commentators, this EU report was fascinating. It seems almost impossible to imagine the EU forcing countries such as Germany or Spain (and others) which have done so well out of FITs, to give them up. That an EU document would provide overwhelming evidence in favour of FITs being more cost-effective than obligations, was unexpected but welcome. That it still maintained the longer-term hope of bringing all the mechanisms together in one pan-euro least-cost, tradable mechanism seems to be expected, given the EU's wider views.

The interaction of country policies, energy agencies and economic regulators

A key theme in this book is that the political paradigm influences both the design of policies and the principles upon which institutions are set up. Another key theme is that a move to a sustainable energy economy requires change across the energy system, not just a focus on technologies or policies of economic regulation. The four countries briefly discussed in this chapter, all have energy agencies and economic energy regulators. This section examines the inter-connections between these institutions and policies in these countries. In brief, those countries, which are 'just do it' countries, ensure that the policies (i.e. FIT) in support of renewables are situated within legislation. The economic regulators in turn are required to work with that legislation. In effect, this means that the economic regulator has to work around legislation which overlaps with their remit, as with priority access of generation; network connection rules, and so on.

Denmark has an energy agency (the Danish Energy Authority) which was set up in 1975 and, since 2005, is an Authority under the Ministry of Transport and Energy.

> It carries out tasks, nationally and internationally, in relation to the production, supply and consumption of energy ... By establishing the correct framework and instruments in the field of energy, it is the task of DEA to ensure security of supply and the responsible development of energy in Denmark from the perspective of the economy, the environment and security ... it is the task of the DEA to advise the Minister ... to administer Danish energy legislation. (www.ens.dk)

The Danish Energy Regulatory Authority (DERA) on the other hand is 'an independent authority engaged in forward looking supervision of monopoly companies in the Danish energy sector'. The members are appointed by the Danish Minister of Transport and Energy for four years but the secretariat is managed by the Danish Competition Authority. The decisions of DERA can be appealed to the Energy Board of Appeal.

Spain has a very devolved energy agency structure. The national energy agency (IDEA, the Spanish Institute for Energy Diversification and Saving) brings together regional and local energy agencies. They act to provide information and advice to their national, regional or local political bodies. The National Energy Commission (CNE) regulates Spain's energy sectors. Its goals are to ensure the existence of effective competition in energy systems and their objective and transparent functioning for the benefit of all operating in those systems. The economic and financial control of CNE is undertaken by the State Controller's Office. The CNE is attached to the Ministry of Industry, Tourism and Commerce. It has to submit annual budgets to the Treasury for approval. Its functions are set out by law but include issuing proposals and reports on energy planning.

The Netherlands has a model closer to the UK's where markets are competitive and energy policies are decided and implemented by the Ministry for Economic Affairs. However delivery is executed through SenterNovem, an energy agency, which has responsibility for policy on innovation, energy and climate change, environment and spatial planning, and where the work on transition management is based. 'SenterNovem promotes sustainable development and innovation, both within the Netherlands and abroad. We aim to achieve tangible results that have a positive effect on the economy and on society as a whole' (www.senternovem.nl). It's mission statement is: 'we aim to implement Government policies on innovation, environment and sustainability in inspiring, forward-looking and professional ways and to promote their coherence. We get cost-effective, quantifiable results, applying the best possible mix of policy instruments based on experience and synergy.'

The German Energy Agency (Dena) was established in 2000 and has more than 100 employees. It is the competence centre for energy efficiency and renewable energies. Its manifold objectives include the rational and thus environmentally friendly production of sustainable energy systems with a greater emphasis on renewable energy sources. To this end, Dena initiates, co-ordinates and implements innovative projects and campaigns at national and international levels. It provides information to end consumers, works with all social groups active in politics and the economy and develops strategies for the future supply of energy. Its shareholders are the Federal Republic of Germany and the KfW Bankengruppe (www.dena.de). Germany has a combined regulator: the Federal Network Agency for Electricity, gas, Telecommunications,

Post and Railway (BNetzA). Their task 'is to provide, by liberalisation and de-regulation, for the further development of their areas of concern. The retail price controls are not part of their remit.' Their legal framework with respect to energy is the Energy Act, which also implements the FIT (www.bundesnetzagentur.de).

The conclusion that this book draws from this gallop across the energy agencies and energy regulators is two-fold:

- that having an energy agency in of itself will not necessarily lead to a sustainable energy future. However, having an organization which has clearly defined legal requirements of it, is useful however since it separates it from political intervention;
- that 'just do it' countries set up environmental outcomes by law and then require the economic regulators to work around that. This means that those economic regulators can get on with their economic regulation without having to worry about balancing environmental concerns.

America

This book does not intend to give an overview of all renewable energy policies in place. However, it cannot resist commentating on the American schemes as compared to the UK's. Although popular in Europe, FIT-type schemes are not ubiquitous. Wind energy has developed quite rapidly in the US under a combination of FITs and state-based competitive quota systems (obligations or Renewable Portfolio Standard, RPS) at state level, although they are both backed up by a federal production tax credit system. The details of the different obligations in the US vary widely. However, the important point of difference is the following: the UK Renewables Obligation is an obligation on suppliers to buy a certain percentage of their supply from renewables, and after that all the details of the contract are between the supplier and the generator. In the US, however, no such mechanism exists. All the obligations in the US are either for a block of capacity (e.g. 2,000 MW) or for a percentage of supply, but all provide for a minimum contract length or price, or both. The UK mechanism continues to be the most risky in place (Wiser et al., 2006; REN21, 2007).

Conclusion

This chapter has not tried to provide a detailed overview of the different sustainable energy policies in place in various countries. What it has tried to do is to pull out certain aspects of those policies to support the broad idea of the book: that the principles of the political paradigm in place in any country will shape the extent to which that country can implement policies which support a move to sustainable energy. If those political paradigms do not enable intervention in energy markets, grid connection rules and costs and risk-free payments then it is likely that those countries will not have successful sustainable energy policies.

The chapter also argues that this will be reflected in a country's innovation policies. That countries which are successful in 'doing' things for sustainable energy are those which are on one side of the innovation fault-line, meaning that they recognize that making the transition to a sustainable energy system is a system issue; is about connectivity; is about supporting non-least-cost policies; accepting that innovation is not linear and predictable; and that they need to reduce the risk of investment.

Finally, the countries which are successful in developing renewables, notably Germany and Spain, implement their sustainable energy policies via legislation which their economic regulators are required to work around. Germany and Spain continue to be successful, dynamic countries despite intervention in support of specific technologies.

9 'Just do it' – Solutions, Opportunities and Realities

This book has not attempted to provide a detailed description of sustainable energy policies, an increasingly urgent topic which has already been covered in many respected books and papers. It has tried to highlight and explain the big blocks, or barriers, to the creation of a sustainable energy system, based on the available evidence of how things work around the globe.

The move from the current energy system to a sustainable one, requires change in all parts of the system. This is with policies from the politicians; within the rules and incentives within economic regulation, altering the planning system, and so on. At root, this book believes that a powerful, interwoven framework of policies has to be put in place to move from here to a sustainable energy system. This requires political decisions and actions, including a recognition of the urgency of the situation.

Its fundamental argument is that the current political paradigm in place in the UK will not help the UK to achieve sustainable development. In order for the UK to achieve a sustainable future, there is a need for a political paradigm shift. The big message for those in Government, civil servants, institutions, companies and individuals, who want a sustainable future, is to start arguing for, or taking action, to increase the pressure for a paradigm shift.

Judge Governments by what they do, not what they say

Because of the process of regulation, and its interwoven nature with wider society, arguing for major change is recognized as a potentially difficult step. Nevertheless, the urgency of climate change and the evidence of the slow rate of change in the sphere of economic regulation, means that radical transformation has to occur, even if that happens in a number of incremental steps. The weight of past technical, economic and policy decisions which have created the current, dominant configuration tend to conspire to make such change difficult. In addition, the high degree of co-dependence of components (between the dominant generating technologies, distribution lines, transmission, markets, business and consumer behaviour, both within the system and between the system and its environment) implies that the trend in any such shift would be gradual, and demonstrate incremental rather than radical innovation. This may eventually realize the policy objective, but is unlikely to be achieved quickly enough to meet the timescales set by politicians to reduce $C0_2$ emissions from the system. This is a fundamental problem for policy makers – they need the rapid development and deployment of both generation and network innovations to achieve their targets, but the long life of existing assets, plus the co-dependence of components mean that the policy measures which will have to be set in place to achieve this must be extraordinarily far-sighted and complex.

As the Stern Review stated: 'climate change will affect the basic elements of life for people around the world – access to water, food production, health and the environment. Hundreds of millions of people could suffer hunger, water shortages and coastal flooding as the world warms' (page vi of Executive Summary, Short) and around 15–40% of species face extinction with 2°C of warming (Part II, ch. 3 p56).

The Stern Review also clearly shows the different effects, and their probabilities of happening, of different parts per million (ppm) $C0_2$ by volume in 2100. It exposes the urgency of having policies in place which achieve 550 ppm by 2100, and preferably 450 ppm. The UK Government says it wants to reduce $C0_2$ by 60 per cent by 2050 from 1990 levels, which at best equates to 550 ppm. If the Government is serious about this, it has to put in place policies which are going to deliver emission reductions, and quickly. So far, this has not been the case. This book has shown that the current political paradigm principles constrain the design of policies, rules and incentives within the energy system, and as a result they either do not work or they do not work as well as they could have done. The Government has to adopt a new set of principles, thereby moving to a new political paradigm and a new set of policies. This chapter sets out what the Government has to do to enable this.

Do we need a paradigm shift or can it happen anyway?

The argument was made in Chapter 3 that the characteristics of a sustainable energy system would be very different from that currently put in place. This author argues that a nuclear future cannot lead us to a sustainable energy future, because it is an electricity-only technology and cannot provide the amount of energy required into the future sustainably, safely, flexibly or efficiently. A sustainable energy system is therefore one based on new low carbon technologies across the energy spectrum – electricity, transport, heat and demand reduction. It will require very different ways of doing things and as such needs substantial amounts of innovation. Chapters 4 and 5 argued that the current paradigm prefers the large, technocratic answers and has not really connected to the requirements (and benefits) of a decentralized system. It has viewed decentralization as having difficulties from its own perspective – for example, it is easier to deal with six large generating companies that interface with 20 million individuals rather than develop an individual-focussed or more decentralized policy. In reality, the large companies can provide only some of the answers (although they can be useful answers) to the problems and individuals have to participate responsibly since their actions are central, arguably the key, to the outcome.

There are a number of reasons why a more decentralized system has benefits:

- Cost – it has been estimated to be far less costly for networks.
- Efficiency – transmission losses etc. are reduced.
- In itself, putting it in place stimulates innovation.
- Allows individuals to participate and to take responsibility for their own actions – if they have their own micro-power; if they buy green electricity; if they have a visible meter.
- Getting in new entrants, which can increase investment, innovation, competition, skills, and so on.
- Accessing new forms of investment capital.
- Ability to diversify forms of technology to different resources which stimulates new diversity – of technology, resource, skills, regional development.

This is not to say that there are not beneficial technical and operational economies of scale for larger systems. This is not a 'small is beautiful' book. It has argued that the current political paradigm is inappropriate for dealing with the challenges of climate change because the paradigm does not feel comfortable with dealing with the more diverse nature and requirements of a sustainable energy system. However, the point is that what is needed is an environment conducive to change and innovation, thereby providing the opportunity to develop in whatever way is appropriate to achieving the desired outcome. The current paradigm imposes constraints for purely economic reasons, often with narrow short-term horizons, which lock-out the smaller or newer actors.

Chapter 3, and to an extent the previous chapter, described what the academic literature says about making that transition from one state to another, from one paradigm to another: how to move from the 'dirty' energy system to a 'cleaner' greener one. The innovation literature is made up of a wide range of groups, individuals and theories. This includes those within the broad transition management world who focus on the interactions between society and technologies and how that society changes. They, broadly, argue that a paradigm (or landscape) is made up of all the interactions and factors that make up that paradigm, such as the political framework, institutions, economic framework, culture, laws, society, and so on. All those factors have to be taken into account when attempting to channel that paradigm in the direction of sustainability (a difficult enough task, as discussed elsewhere). Economics is therefore only one part of the jigsaw puzzle of change. Moreover, other academics or academic groups fit within this broad paradigm approach. For example, the work of Tim Foxen, who recognizes the transition management literature but who focusses more on the early needs of technology innovation. Similarly, the work of Adrian Smith, on the development of niches, can be seen as an important part of the system change literature. Others, such as Elizabeth Shove, argue that innovation is inherently uncertain and the notion of 'transition management' should be treated with great care since a desired outcome cannot be achieved simply by putting a policy (however broad and joined up) in place. Any outcome, desired or otherwise, will be the result of a huge number of factors.

Leadership, governance and connectivity

Governing is complex; policies do have to work together. Governments do therefore have to have 'principles' of government as a way to ensure 'good' governance. The problem for Government is cascading the principles out to widely differing policy demands, in a complex world. That complex world is interconnected by countries, trade, politics, pollution, security, equity and people issues. Climate change is a problem of a different scale from those that Governments have dealt with before; and the success in overcoming it also involves a very different mix of scales of technologies and actors. Diversity itself adds new demands of communications, which can bring benefits and opportunities. And the urgency of it is new. And all of this is combined with uncertainty about how to bring about the transition to a sustainable energy system, to create a brave new world of environmental stewardship. We are not sure which technologies are the right technologies; we are not sure how to involve individuals and appropriately alter human behavioural issues; we are not sure what the relationships and roles should be between countries; and between Government, regulators, companies and individuals.

This book recognizes and accepts all of the above. But it is fundamentally a book about policy, application and reaching certain outcomes. While it recognizes there is no clear link between policy and outcome, it does not accept that no policy is better than any policy, although it is happy to accept that sometimes no policy is better than a particular policy.

Practical application

The political paradigm shapes the principles from which Government policies are derived and how institutions are set up, and the rules and incentives which emanate from them. Regime or system change requires innovation across the spectrum of actors within the paradigm. This book has tried to bridge the gap between applied policy development and 'theory' in the widest sense. It feels comfortable in 'knowing' what works 'in general' in terms of policy, while always recognizing that innovation is not linear or predictable; that every situation is different; and, therefore, there is never any guarantee of successful policy transferability from one situation to another. From a policy perspective, the choice faced tends to be between not doing anything (as the best option based on evidence); doing something positive; or doing what is possible, given any particular situation. It is very rare that a 'perfect' policy can be designed and implemented, even if there were one (in itself a highly contested idea). Generally, policies are implemented on the back of old ones.

With this in mind, this final chapter sets out what the author sees as necessary steps to deliver a system (or regime) change which enables a sustainable future. At root, it argues that the power (meaning, in conventional terms, the ability of an individual, group or institution to do something, or cause something to happen) of the current political paradigm has to be broken by establishing an environment which is more powerful, thereby enabling it to be more supportive. This book has talked about the 'band of iron', meaning the fundamental power of the current political paradigm, embodied in its principles which cascade down through its institutions, policies, rules and incentives. It is this band of iron which has to be broken. This band of iron is the rules and incentives which dictate how the current energy industry companies make their money. Unless it is broken, the current energy companies will continue to make their money from the same fundamentally non-sustainable practices; the UK will continue to implement sustainable energy policies destined for failure; have a non-participatory energy system dominated by a few, in-control, large companies; and have a limited amount of innovation, which becomes ever more 'closed down' rather than 'opened up'.

The way forward (a third way?)

Geel's paradigm model has three levels: the landscape (i.e. a country and its political framework and principles), the regime (i.e. the current energy system) and the niches, some of which develop and make it into the regime and others which collapse. Society, technologies and so on drive the paradigm, but are also the product of the paradigm. The paradigm is therefore a mass of inter-linked factors, whether technological, economic, political, social, and so on. This implies that regime change has to be gradual. If any one element is changed too rapidly on its own, you may end up with unhelpful overall system disruptions, although some argue that a degree of disruption is necessary to stimulate change (Christensen, 1997, 2003; Christensen et al., 2004). Even so, the other elements will have to catch up and change, if there is to be overall change.

The question is whether gradual regime change means that incremental change on the part of the energy system actors is the only way forward, and whether it will lead to regime change. Will gradual change break the power of the band of iron, or can the rigidity of the band only be broken by a one-off powerful, individual exertion of strength – and across the board – and if so how should this come about?

In essence, how should this new political paradigm come about when the current political paradigm is not supportive of it? This author acknowledges gladly that the UK Government wants to meet the challenges of climate change, but argues that it is putting in place policies, rules and incentives which are not working. This book puts forward recommendations of what Government should do to make them work. If Government is serious about the problem, it will make the necessary changes. A paradigm will change when society is ready for it to change. All those who push for change in their way (whether, for example, writing a book, signing up for a green electricity tariff or jostling for a new sustainable energy agency) are increasing the pressure for a new political paradigm able to deal with the complexities and difficulties of climate change.

Political paradigm change

The implication of a paradigm is that change has to be gradual because so much has to alter, in terms of depth and reach, to achieve a permanent shift in the character of the paradigm. The two political paradigms of the twentieth century were the Keynesian Interventionist State and the Regulatory State, ushered in by Margaret Thatcher. New Labour, originally led by Tony Blair, has not fundamentally altered the political paradigm, even if their goals have been somewhat different from those of Margaret Thatcher.

To a degree, a change in political paradigm ushers in a time of wider societal change, even if that change takes time to bed in and settle down. For example, Thatcherism arrived and led to a series of social effects such as the miners' strike in 1984, the poll tax riots, and so on. Nevertheless, the voting in of Margaret Thatcher led to a change which was a breakthrough in the paradigm – it imposed a rupture – even if it only came about as a result of a long build up of pressure on the band of iron beforehand, and the time taken to bed down afterwards. At some point, the pressure on the band of iron causes it to snap. All those ways of being that it held in place, are then free to move, drop away, reform, be added to until the new paradigm coalesces again into a new band of iron. Changing a paradigm is fundamentally a combination of incremental change. An event may cause it to break, but this is in effect the 'straw on the camel's back' which precipitates a more general change.

Importance of participation

All actors, whether politicians, civil servants, regional and local authorities, companies, NGOs, communities, individuals, global actors (and so on) involved in the build up of pressure are therefore central to a paradigm shift. This highlights the importance of individuals, as voters. Every few years there is a very obvious expression of voter concern. Every few decades, their concerns appear to come together in 'landslide' defeats or victories. Society collectively comes together to express a view and paradigm shifts seem linked to this. A democratic society is one with a leadership which attempts to keep society (of all levels and in all forms) informed and involved in decision-making. Governments which endeavour to do this are more likely to have an informed society. Equally, that society is more likely to have views on what that Government is doing. From the perspective of Government, there is always the temptation to reduce transparency, to close down participation, to impose change.

A parallel to this within innovation, is the extent to which Governments, institutions, local authorities and so on are able to encourage sustainable innovation, as opposed to no innovation or the wrong sort of innovation. Given the concerns above of the uncertainty of how innovation occurs; and the questions of whether paradigm shift occurs as a result of incremental rather than radical change; an important element in trying to encourage an environment conducive to change or innovation is the effort made at 'opening up' rather than 'closing down'. This is across the system. In this case, for example, a FIT would be chosen rather than an obligation. Given an active choice, the choice which better 'opens up' should be taken rather than 'closing down'.

If the ethos of society is to 'open up' then there is more chance of innovation happening. Similarly, where possible, enabling participation is a way to enable connection between choices, thereby enabling a more informed means of that choice. In the same way, where decisions are being made that can enable participation to a greater or lesser degree then the decision, or principle, in support of participation should be taken. This has to become an active decision of Government, and is likely to occur as the pressure for it becomes so great (from individuals, companies, etc.) that political leaders take that step.

Balancing incremental change

Originally, the regulatory state paradigm was determined to inject greater economic efficiency into the ex-monopoly companies and monolithic state organizations of the UK. As time has gone on, the issues that the energy regulator has to deal with have altered in character and urgency. The inability of the economic regulator to incorporate qualitative, indirect values, such as diversity, can be seen as a central flaw in dealing with these new issues.

The economic design of Government policies and the rules and incentives of economic regulation tends to lead to incremental change. As this book has shown, these policies, rules and incentives have not necessarily worked well in developing a sustainable energy system; they have sometimes been more expensive (Renewables Obligation), been unsuccessful (innovation in distribution networks), led to a long drawn-out and uncertain process which finally came back to a regulated non-market approach (offshore transmission), or are undermining (the electricity market). One way through this is to say 'just do it'. The point of these examples is that their design is based on the principles that policies should be technology and fuel blind and/or where possible based on competition. The UK's political paradigm supports a de facto hegemony of economic design. This book is arguing for a middle way, which is that sometimes a policy, or rule or incentive is targeted for a specific technology. This middle way can be viewed as more complex to operationalize than the simple, clear principle that wherever possible competitive means are used.

The argument put forward in this book is that to reach a sustainable solution for the country it is necessary to break the band of iron, and that will require a shift in the political paradigm to one where the principle of economic dominance is diluted so that, in matters of climate change, the environmental options may take precedent. That means as far as possible, when designing a policy, the economic dimension which slows the process down or limits innovation or change should be by-passed. The Government has to move to a principle in matters of climate change to ensure a successful delivery of the policy. In effect, this means moving to a determination that the sustainability has to be achieved at whatever the cost, rather than the current possible sustainability from the desired economic expenditure.

As argued earlier, and in Chapters 5 and 8, these 'just do it' policies do not necessarily mean that they are more expensive or less effective. A key argument is that the costs of one energy system versus another are hugely contested. For example, is a nuclear energy system more or less expensive than a decentralized one? What is the basis of such a judgement? What (indirect) benefits are included and how are they valued?

As Paul Ekins has so clearly stated:

> the fact is that, in market economies, the kind of structural economic reform that will be necessary to address climate change is nearly impossible if it is working against market signals. If it relies only on other policy instruments it will be costly and probably ineffective. Not only is the price mechanism essential for resource reduction but price signals also increase the impact of other instruments, such as information and voluntary agreements. (Ekins, 2006)

This book agrees. The UK is a market economy and will remain so, as will most countries of the globe. Ekins puts forward recommendations for how prices within the market economies can start 'pointing in the right direction'. He argues for addressing market failures through environmental policy which he says is made up of market-based instruments, environmental tax reform, regulatory instruments, voluntary agreements and information-based instruments (e.g. eco-labels).

However, this book has argued that the economic sphere is only one area of the wider paradigm. That whole paradigm has to move over to addressing climate change. Successfully undertaking economic structural reform is one important area but stimulating innovation and technology development, at the rate required to meet the global environmental imperatives, is another. Instituting policies, rules or incentives, in certain clear situations, for certain outcomes, which 'intervene' in a market or require policies, rules and incentives which target particular technologies should be undertaken in parallel to the structural economic reform. To do this is not to undermine or undervalue the importance of structural economic reform. It recognizes that while economics is one, albeit important, sphere of the paradigm, all spheres should be doing their 'bit' to structurally change for a sustainable future. Innovation and technology development are equally as important. At some point, hopefully, in the future the structural reform of the paradigm will have been successful and a market economy of a sustainable energy economy will continue.

'Just do it' countries

As the previous chapter showed, some countries have 'just done it' meaning: that they have put in place policies which reduce the risk of investment; increase the confidence of the investors in the Government's determination to support sustainable energy over the long term; and make it easy for sustainable energy developments to happen. They have a determination to deliver change: they want to make sustainability happen, at the lowest cost (defined in a long-term, qualitative and quantitative sense) or whatever it costs, rather than deciding on a cost and seeing how much sustainability happens. The UK has to switch to the former from the latter viewpoint. These 'just do it' policies tend to be narrow and focussed on an outcome. It does not preclude a fundamental support of market mechanisms for ordering their society's choices and, on the whole, they undertake them in parallel to economic structural reform, argued for in the section above. However, these countries are prepared to target particular technologies with policies which intervene in their electricity market (for example, by requiring priority access for the generation) or which have clear, not necessarily cost-reflective rules for access and connection to the grid. These countries have not thrown economic sense 'to the wind' or undertaken 'foolish' actions. They have undertaken well thought out innovation policies to support the development of sustainable technologies, for a number of reasons, in parallel to supporting their own competitiveness and economic well being.

A powerful force

The UK could institute such a policy framework which exhibits the determination to deliver sustainability. A powerful, interwoven framework of policies has to be put in place to move from the current situation to a sustainable energy system. It is a political decision and it requires a political determination to turn it into action. This does not need an energy agency, or any other bodies. It could be undertaken by a determined Government through legislation. Moreover, it should be cross-party and it should ensure continuity so that future politicians could not easily rescind it.

Legislation for sustainable energy

As the previous chapter concluded, the countries which effectively 'just do it' are those which have been successful in sustainable energy deployment (e.g. Denmark, Spain and Germany) in part by revising approaches to short-term economics, by adopting positive innovatory strategies and mechanisms; and by having regulatory agencies which are required to achieve defined outcomes. None of the energy agencies or regulators discussed in the previous chapter have the type of independence which occurs in the UK. In all those countries, decisions about desired outcomes are made by Government through Parliament and the agencies or regulators become executives of those decisions. Those decisions may have been informed by their analysis; but nevertheless, in the end they are political decisions.

This is what is needed. At root, this means establishing legislation which implements the principle that, in certain circumstances, it is not just acceptable, but is appropriate, to implement a policy to ensure an environmental outcome.

This book is arguing, in effect, that the hegemony of the economic goal of the Government's energy policy is inappropriate in all circumstances, since the vital and unyielding nature of energy policy is to move to sustainability (while at the same time being secure).

Primacy of intervention in environmental matters

Establishing a clear framework of intervention for environmental objectives is not such a great jump to make, although clearly it has to be implemented within clear parameters. As mentioned above, this is effectively what the majority of European countries do. One aspect of Ofgem the regulator, is that it has this sense of its own power because of its independence and its duties. It does not feel the need, rightly, to do what any Government department would like it to do. It does what it wants, within its duties. The Environment Agency is a less divisive example. It has the power to prosecute companies which exceed their permitted pollution. The Sustainable Development Commission also has certain powers to ensure that Government departments fulfil their sustainable development remit. The Climate Change Committee has so far been a powerful force for the good of the environment – but it is not allowed to make recommendations for policy only to explain the science behind climate change and what this means for the amount of emission reduction the UK should be achieving.

The point is that the UK does not have a political paradigm which includes an institution which has the power to ensure that the environmental imperative is taken seriously, and that the Government can 'just do it'. This is what is needed. Without this, the UK will not be playing its part in the global environmental imperative; nor will it be able to prosper from the opportunities such an imperative implies. This is not the same thing as calling for 'cross-party' consensus on climate change policy, as has been proposed by the All Party Parliamentary Climate Change group, although that may be worthwhile, since it could lead to clear and more widely accepted policy recommendations.

A central aim of this book has been to show that society's paradigm (not simply the political paradigm which is one part of it) is inter-linked, and is far bigger than the economic aspects of it. No doubt there are those who will raise their eyebrows at the 'just do it' 'just get on with it' view to policy. It could be taken as naïve (everything costs something and is central to the reality of politics and policy-making). No doubt it will be argued, and to a degree it is correct, that rigorous, transparent, democratic policy-making should include cost-benefit analyses, regulatory impact assessments and least-cost this and that. However, this should not be the sum of policy-making. It is not that this book disagrees that the cost of policies should be disregarded. What this book has argued though is that the narrow economic basis of policy design is not working (either in an absolute sense or because the policies are more expensive or less effective than non-market approaches) with respect to sustainability, and arguably for security. Certain policies, rules or incentives may, theoretically, be more economically efficient. But if a policy, rule or incentive is not working (in terms of delivery of what it is meant to) and, if that is to a large extent because of the economic design of it, *and* if it is making the desired outcomes worse (i.e. one step back with NETA or BETTA) or less likely to happen (i.e. by increasing risk, by increasing complexity, by having a non-dynamic and non-innovative attitude to innovation; by having a short- rather than long-term vision to a long-term question) then the time has come when another way forward has to be considered.

Economics will be central to any way forward, as argued above. In general, in market economies, price is the most used basis of choice. But in a complex world, where decisions are often made from the rational perspective of the individual (or irrational from the point of economics); where there are market failures; and where factors other than economics are important to outcomes, then other means of stimulating technology development or a paradigm (structural) shift to a sustainable energy future are not only necessary but appropriate.

The opposite of command and control

It is important to recognize the 'just do it' and 'keep it open' view of policy-making is not a step back to command and control regulation. 'Just do it' as compared to the 'economic design' of policies, regulation and incentives is a different approach to getting to the same desired position. One argument is that large companies are central to the movement to a sustainable energy system because it is in their interests to maintain their position. And this of course is true to a large degree, but as discussed in Chapter 2, it is also in their interests to traverse that course at the rate they wish. A competitive-based policy will also complement economies of scale and companies able to access cheaper costs of capital. A mechanism which 'opens up' the market should not undermine the large companies because they should be able to compete in an open market, even if they prefer a more constrained one. What is more of a problem for large companies, is mechanisms which enable entrants in a non-competitive manner, for example as the FIT does. These non-market mechanisms which reduce risk and increase investor certainty which are valuable for innovation and technology development (UKERC, 2007) are exactly those not preferred by the large companies.

However, the basis of the 'just do it' school of policy design and economic regulation is reducing risk; making network access easier; investment easier. All of these aspects of how technologies and societies develop are about increasing participation, new entrants, and so on. They are no more command and control however than the Renewables Obligation or any other of the economic design mechanisms.

It is a system that enables new entrants, and thereby, in some senses, opens up competition. It is about ensuring that the momentum of the old system does not continue to exclude new ways of doing things. And it does so, not by favouring one company over another but simply by equalizing the incentive given. Large and small companies can equally benefit from the FIT or from clear access rules. What it is doing is lowering the hurdles or barriers for new companies or technologies to enter into the new energy system.

Steps to take

Other countries have been more successful in developing sustainable energy policies. All countries are different; none are perfect; and all policies have to be country specific. However, the UK has a poor record compared to most European countries. This book argues that this is because the UK's political paradigm constrains and channels the design of its policies and its rules and incentives within its narrow view of economic regulation, *and* because these constraints directly affect how successful policies are or how successful economic regulation is in encouraging sustainable energy. The UK needs new policies combined with new rules and incentives within its economic regulation. The Government says it wants to meet the challenge of climate change. This book argues that, if this is to be the case, then it needs new principles; and for this to occur, and for them to be implemented, the character of the political paradigm has to change. Only then can the principles be loosened enough so that the policies, rules and incentives within the energy system are altered to become conducive to innovation.

As set out in Chapters 1 and 2, the Government has to move to the right side of the innovation fault-line, which was set out in a general way as:

- having an understanding of what 'innovation' is, and a recognition that not all of it is 'good' and therefore it has to be directed;
- accepting that markets are not always the best way forward for making all choices – although certainly they are for many decisions

(if not the majority) and will continue to be central to any future sustainable energy system;

- accepting that 'picking' a technology to support is not only acceptable but necessary to 'channel' innovation policies;
- accepting that choosing to support an environmental option, which may not be a short-term least-cost measure, rather than choosing the economic or market option, may be appropriate and necessary and provide a great deal of additional value, albeit not in a way which can be valued monetarily;
- accepting that trying to meet the challenges of climate change is a 'system' issue not a technological-only issue.

While trying to transform the Government's views on innovation, the parallel steps that need to be taken are:

- The power of the political paradigm (i.e. the band of iron) needs to be broken by increasing the pressure on the principles of the paradigm to change to those set out below, until such a time that it breaks.
- In matters of climate change, the environmental option needs to take precedent (and this may require the establishment of a sustainable energy agency with the power to see this through).
- The goal of the political paradigm has to shift to achieving sustainability (at lowest cost but defined in a long-term sense which includes qualitative factors) rather than establishing a cost and seeing how much sustainability comes out (as it is now).
- The UK has to become a country which recognizes the urgency of climate change and 'just does it', as the bullet above indicates.

Given that paradigms cannot be changed at will, it also requires a number of incremental changes, stimulated by those able to make those changes, which in turn requires new political paradigm principles from which policies, institutions, rules and incentives emanate:

- Governments, and all other institutions, companies and individuals, should endeavour to create an environment conducive to innovation, which means: implementing policies, rules and incentives which reduce investment risk (i.e. making it easy for investors to invest, to connect, to raise capital, to become involved and take responsibility); increase the certainty of long-term political will in support of achieving sustainability by implementing targets and so on; stimulating new entrants; encouraging participation; enabling an opening up of options and participation rather than a closing down.

- Governments, and all other institutions, companies and where possible individuals, should take a longer-term view to their actions and assess costs and benefits in terms of both qualitative and quantitative values.
- Governments, and all other institutions, companies and individuals, should overtly support a system approach to innovation by moving away from the hegemony of economic design within the policy and economic regulation arena.
- Governments, and all other institutions, companies and individuals, should accept that while structural economic change for sustainability is vital, so are the other spheres of the paradigm.
- Governments should build flexibility into their policies and institutions, through assessment.
- Finally, all actors, whether to do with planning permission, grid access, policy development, are important and should play their part. Part of the opening up of innovation policy is enabling access, participation and involvement.

In particular, this book argues that becoming a just-do-it country is a political choice, The UK has to move from its current position where the 'agency' of moving has been given elsewhere, precisely so that politics cannot 'interfere' in policy procedures. Hence, Ofgem's independence and the privatization of the energy industries. While valid at the time, this arms-length decision-making is no longer suitable because of the extraordinary demands of climate change mitigation, and increasingly energy security. It is inappropriate for any other body other than Government to take these far-reaching decisions.

The Climate Change Committee already gives an annual report about the state of the climate change emissions. The Government should alter the duties on the energy regulator, so that the regulator alters its focus to delivering a sustainable and secure energy system. The Government should dispense with 'guidance' to the Regulator but issue requirements, based on the Climate Change Committee reports.

Conclusion

Political paradigms do not just change. They evolve in response to pressure. This book is arguing for a new way of designing policies and incorporating sustainable innovation evidence. This author argues that in matters of climate change, the environmental choice should, in some situations, take precedent. This should be a political decision and be implemented through clear, legislated action. This would enable a powerful, interwoven framework which links policies, innovation, economic regulation, planning, consumption and technology issues to move to a sustainable energy economy. It will require intervention in support of these sustainable technologies within markets and within economic regulation. This does not preclude dynamic, competitive markets. It should be viewed as a necessary reduction in hurdles to enable the system-wide transformational forces to take place. Economic regulation would continue, but would be secondary in matters related to climate change. There are many examples of this in Europe.

The push has to come from the Government; only through its determination and legislation can confidence be instilled throughout the energy system, thereby stimulating the necessary investment and participation from all quarters. Other countries do this in parallel to successful economies. It is time that the UK started to do the same. This is not arguing that economic theory and competitiveness be sidelined. However, it is arguing that innovation and the importance of sustainability take a rightful place beside them.

The key to achieving this is to understand how a paradigm can be changed, particularly when its power is as inter-linked and all encompassing as it currently is. This book has rather lamely ended by arguing that individuals, whether as Ministers or schoolchildren, all try to stimulate change towards a sustainable future. However, the essence of the next paradigm has to be a determination to act in a sustainable manner in every sphere of life, both domestically and globally. This seems far off at the moment, yet the extent of the climate challenge requires this.

Notes

1 Breaking Free of the Band of Iron

1. The IPCC classifies its level of confidence in the link between human activities and the observed warming as 'very high', which equates to a 9 out of 10 chance. This is the strongest statement yet linking anthropogenic greenhouse gas emissions and climate warming.
2. It appears increasingly unlikely that this target will be met (DTI, 2006a).
3. 'Eligible' renewables include wind, wave, tidal, biomass and solar technologies, and some hydro power. Energy from waste is not eligible. Output is supported by the Renewables Obligation, which requires electricity suppliers to buy an increasing amount of renewables generation each year.
4. This combines the Renewables Obligation for England and Wales, and for Scotland. The Obligation level for England and Wales in 2005–06 was set at 16,175,906 MWh, and for Scotland was 1,648,679 MWh. The proportion met by Renewables Obligation Certificates was 76 per cent in England and Wales, and 68 per cent in Scotland (Ofgem, 2007a).
5. I write about the principles of the regulatory state paradigm to reflect the framework and constraints of Government policy development. The RSP, as with all political paradigms and as discussed in greater detail in the next chapter, combines two strands: the political framework and its institutions, which are set up according to principles (which derive from the political framework) and which tend to be the executor of policies. The political framework (meaning the Prime Minister, the Cabinet, departments, Cabinet Ministers, MPs, Select Committees, legislative authority and so on but also the wider underlying political context of society) is not a monolithic body. While I talk of 'the principles of the RSP' I recognize that it is a very broad statement and there is within that paradigm a spectrum of views, some of which will be utterly opposed to Government policies which have been designed in a particular way because of those principles. However, to explain the context of each reference to the 'political paradigm' or the RSP would not lend itself to a flowing narrative.

4 Preferable Intervention – the Pursuit of Nuclear Power

1. Several other Magnox stations have already closed down and are undergoing decommissioning.
2. The Nuclear Decommissioning Authority is a non-departmental public body established in April 2005 to decommission 20 nuclear sites previously owned by BNFL, the UK Atomic Energy Authority and Ministry of Defence sites. It does not own British Energy's sites, although any decommissioning strategy developed by BE has to be approved by the NDA, and it oversees the liabilities of the Nuclear Liabilities Fund.

3. The Magnoxes were withdrawn first, in July 1989. The AGRs and Sizewell B were withdrawn in November 1989. For a more detailed examination of this period, see Mackerron (1996).
4. The public inquiry into Hinkley Point C finished in 1989, and the station was ultimately given planning permission, but was never built because of the moratorium.
5. Other notable examples are the failed attempts in 1983 and 1987 to find suitable sites. These are set out in more detail in the House of Lords report (House of Lords Science and Technology Select Committee, 1999). The search to find a site for a HLW dump was abandoned in 1981 and has not seriously been pursued since. All of these attempts had met with sustained public opposition.
6. Finland decided in 2002 to build a new nuclear power plant: Olkiluoto. It is being built for the Finnish power company TVO by a consortium. However, having started construction in 2005, the power plant is already two years behind schedule. For details see the website of the Embassy of Finland, London: www.finemb.org.uk/Public/Print.aspx?contentid=100853&nodeid=35864&culture=en-US&contentlan=2.

6 Markets and Networks – Pure Paradigm and Effect

1. Some of the Round 2 sites are more than 12 nautical miles out to sea, and therefore outside the UK's territorial limit. In order to have the legal powers to license and consent wind projects in this area, the Government legislated to create Renewable Energy Zones (REZs) into which it can extend the Section 36 consenting regime. This power is contained in the Energy Act.

7 New Zealand as a Case Study

1. This takes account of blended capacity factors for wind, hydro and geothermal.
2. Huntly extension (e3p), gas and 365 MW; Southdown expansion, gas, 45 MW; compared to White Hill Southland, wind, 58 MW; Te Rere Hau Tararua, wind, 48.5 MW; Deep Stream, hydro, 4 MW; three geothermal enhancements, 58 MW in total at Wairakei, Poihipi Road and Ohaaki; and Manapouri upgrade, hydro, 16 MW. www.med.govt.nz/templates/MultipageDocumentTOC____24880.aspx.
3. Meridian and Contact. Contact announced NZ$2 billion of renewables investment by 2014.
4. www.scoop.co.nz/stories/BU0702/S00353.htm.
5. www.nzherald.co.nz/category/story.cfm?c_id=37&objectid=10430099.

References and Further Reading

Agterbosch, S., P. Glasbergen and W.J.V. Vermeulen (2007). Social barriers in wind power implementation in the Netherlands: perceptions of wind power entrepreneurs and local civil servants of institutional and social conditions in realizing wind power projects. *Renewable and Sustainable Energy Reviews*, 11(6): 1025–55.

Agterbosch, S., W. Vermeulen and P. Glasbergen (2004). Implementation of wind energy in the Netherlands: the importance of the social-institutional setting. *Energy Policy*, 32(18): 2049–66.

Anderson, D. (2003). DTI Economics Paper 4. *Options for a Low Carbon Future.*

Arrow, K. (1962). Economic welfare and the allocation of resources for invention. In R. Nelson (ed.), *The Rate and Direction of Inventive Activity: the Economics of Technological Change* (Harmondsworth: Pelican, 1974).

Arthur, W.B. (1989). Competing technologies, increasing returns, and lock-in by historical events. *The Economic Journal*, 99: 116–31.

Aubert, P.J. (2007). *Energy Transition – the Dutch approach.* KSI Winterschool 2007. Vught (Netherlands).

Ayres, I. and J. Braithwaite (1992). *Responsive Regulation: Transcending the Deregulation Debate* (Oxford: Oxford University Press).

Baldwin, R. and M. Cave (1999). *Understanding Regulation: Theory, Strategy and Practice* (Oxford: Oxford University Press).

Bank of England (2006). www.bankofengland.co.uk/monetarypolicy/overview.htm.

Berkhout, F. (2002). Technological regimes, path dependency and the environment. *Global Environmental Change*, 12(1): 1–4.

Blair, T. (2004). Speech at the launch of the Climate Group, 27 April. www.number10.gov.uk/output/page5716.asp.

—— (2005). Reported by the *Guardian* newspaper, 22 November. Blair says the 'facts have changed on nuclear power'. www.politics.guardian.co.uk/homeaffairs/story/0,11026,1648182,00.html.

—— (2006a). Reported in *Daily Mail* newspaper, 30 November. Blair's £150-a-year nuclear power tax. www.dailymail.co.uk/pages/live/articles/news/news.html?in_article_id=370140&in_page_id=1770.

—— (2006b). New Zealand Climate Change Speech, 29 March. www.number10.gov.uk/output/Page9260.asp.

—— (2006c). Speech to the CBI Annual Dinner, 16 May. www.number10.gov.uk/output/Page9470.asp.

BMU (2007). Development of Renewable Energy Sources in Germany in 2006, Graphics and Tables. Version: June 2007. www.bmu.de/files/pdfs/allgemein/application/vnd.ms-powerpoint/ee_zahlen_2006_en_ppt.ppt#256.

BNFL (2001). Submission to the Performance and Innovation Unit's Review of UK Energy Policy.

—— (2005). Submission to House of Commons Environmental Audit Committee Inquiry. Keeping the Lights On: Nuclear, Renewables and Climate Change.

—— (2006). BNFL Submission to Energy Consultation, March. www.bnfl.com/UserFiles/File/BNFL%20submission%20to%20DTI%20Energy%20Review%202006.pdf.

Botting, D. (2005). Technical Architecture – a First Report. *IEE Power Systems and Equipment Professional Network*, Issue 1.3.

British Energy (2001). Replace Nuclear With Nuclear. Submission to the Government's Review of Energy Policy.

—— (2006). Submission by British Energy Group plc to the Energy Review, April. www.british-energy.com/opendocument.php?did=394.

Buen, J. (2006). Danish and Norwegian wind industry: the relationship between policy instruments, innovation and diffusion. *Energy Policy*, 34: 3887–97.

BWEA website: www.bwea.com/offshore/index.html.

Carbon Trust (2006). Policy Frameworks for Renewables – an analysis of policy frameworks to drive future investment in near or long term power in the UK. July. www.carbontrust.co.uk/Publicsites/cScape.CT.PublicationsOrdering/PublicationAudit.aspx?id=CTC610.

Central Networks (2005). Central Networks leads industry on developing innovative solutions to meet government green targets. Press release, 29 June.

Chapman, R. (2004). The Dutch Sustainable Energy Transition and its applicability to New Zealand. *Public Sector*, 27(4): 23–8. www.ipanz.org.nz/SITE_Default/x-files/10202.pdf.

—— (2006). A way forward on climate policy for New Zealand. Paper presented at Climate Change and Governance Conference, Wellington, March.

Chapman, R. and J. Boston (2006). Introduction: critical issues. In R. Chapman, J. Boston and M. Schwass (eds), *Confronting Climate Change: Critical Issues for New Zealand* (Wellington: Victoria University Press). November.

Christensen, Clayton M. (1997). *The Innovator's Dilemma: When New Technologies Cause Great Firms to Fail* (Cambridge, MA: Harvard Business School Press).

—— (2003). *The Innovator's Solution: Creating and Sustaining Successful Growth* (Cambridge, MA: Harvard Business School Press).

Christensen, Clayton M. et al. (2004). *Seeing What's Next: Using the Theories of Innovation to Predict Industry Change* (Cambridge, MA: Harvard Business School Press).

Clark, H. (2007a). Prime Minister's Statement to Parliament, 13 February 2007. Press release, available from www.beehive.govt.nz/Print/PrintDocument.aspx?DocumentID=28357.

—— (2007b). Address to NZ Ambassador's Reception at Fairmont Olympic Hotel, Seattle, Friday 23 March 2006.

Clark, N. (1990). *Evolutionary Theory in Economic Thought* (London: Pinter).

Committee on Radioactive Waste Management (CORWM) (2006). Draft Recommendations. www.corwm.org.uk/pdf/None%20-%20CoRWMs%20Draft%20Recommendations%2027%20April.pdf.

Confederation of British Industry (CBI) (2005). CBI, Powering the Future: enabling the UK energy market to deliver. 21 November. www.cbi.org.uk/ndbs/positiondoc.nsf/81e68789766d775d8025672a005601aa/9a6f3c35b2a6835e802570c0004ec667?OpenDocument.

—— (2006). Government Must Clarify Longer-Term Carbon Plans So Low-Emission Energy Generation Can Compete. Press release, 24 April. www.cbi.org.uk/ndbs/press.nsf/0/841a83040307d48080257155002eeed5?OpenDocument.

Dale, L. et al. (2004). Total cost estimates for large scale wind scenarios in the UK. *Energy Policy*, 32(17). Special Edition: Energy Policy for a Sustainable Future. Ed. C. Mitchell.

Davies, A. (1996). Innovation in large technical systems: the case of telecommunications. *Industrial and Corporate Change*, 5(4): 1143–80.

Davies, C. (2005). Regulatory and License Issues, BWEA Conference on Offshore Wind. Available from www.bwea.com.

Davies, C. and J. Ward (2005). Offshore Networks – possible approaches to licensing and regulation. April. Available from CMUR, University of Warwick website: www.wbs.ac.uk/cmur.

Defra (2001). Managing Radioactive Waste Safely, Proposals for developing a policy for managing solid radioactive waste in the UK. September. www.defra.gov.uk/environment/consult/radwaste/pdf/radwaste.pdf.

—— (2006a). Greenhouse Gas Policy Evaluation and Appraisal in Government Departments. April. www.defra.gov.uk/environment/climatechange/uk/ukccp/pdf/greengas-policyevaluation.pdf.

—— (2006b). Climate Change Legislation. Press release, 30 October, Ref. 463/06.

Defra/Treasury (2002). Estimating the Social Costs of Carbon Emissions. www.hm-treasury.gov.uk/media/209/60/SCC.pdf.

Dinica, V. (2006). Support systems for the diffusion of renewable energy technologies – an investor perspective. *Energy Policy*, 34: 461–80.

Distributed Generation Co-ordinating Group (2005). Third Annual Report 2004/05. March.

DTI (2001). Embedded Generation Working Group, Report into Network Access Issues. January.

—— (2002). Future Offshore: a Strategic Framework for the Offshore Wind Industry. URN 02/1327. www.dti.gov.uk.

—— (2003). Energy White Paper: Our Energy Future – Creating a Low Carbon Economy. Cm 5761. http://reporting.dti.gov.uk/cgi-bin/rr.cgi/www.dti.gov.uk/files/file10719.pdf.

—— (2005). Energy Review – a Secure and Clean Energy Future. P/2005/378. 29 November. www.gnn.gov.uk/environment/detail.asp?ReleaseID=179546&NewsAreaID=2.

—— (2006a). The Energy Challenge, Energy Review Report 2006. Cm 6887. URN 06/1576. www.dti.gov.uk/files/file31890.pdf.

—— (2006b). Our Energy Challenge, Power From the People; Microgeneration Strategy. March. URN 06/993. www.dti.gov.uk/files/file27575.pdf.

—— (2006c). Regulation of Offshore Electricity Transmission, Extension of the GB System Operator Role Offshore. August. www.dti.gov.uk/files/file32874.doc.

—— (2006d). UK Energy and CO_2 Emissions Projections, Updated Projections to 2020. February. www.dti.gov.uk/energy/environment/projections/recent/page26391.html.

—— (2006e). Energy Trends. March. www.dti.gov.uk/files/file27084.pdf.

—— (2006f). Call for Evidence for the Review of Barriers and Incentives to Distributed Electricity Generation Including Combined Heat and Power. www.dti.gov.uk/files/file35026.pdf.

—— (2006g). Regulation of Offshore Electricity Transmission, Government Response to the Joint DTI/Ofgem Public Consultation. March. www.dti.gov.uk.

—— (2006h). Licensing Offshore Electricity Transmission – A Joint Ofgem/DTI Consultation, Partial Regulatory Impact Assessment. November. www.dti.gov.uk.

—— (2007a). Energy Trends. June. www.dti.gov.uk/files/file30881.pdf.

—— (2007b). Meeting the Energy Challenge, A White Paper on Energy, Cm 7124, May, www.dti.gov.uk.

DTI Centre for Distributed Generation and Sustainable Electrical Energy (DGSEE) (2004). Electricity Distribution Price Control Review, Response to the Second Ofgem Consultation. www.ofgem.gov.uk/temp/ofgem/cache/cmsattach/6114_DGSEEresponse_17103_document.pdf?wtfrom=/ofgem/work/index.jsp%C2%A7ion=/areasofwork/distpricecontrol/.

—— (2007a). Transmission Investment, Access and Pricing in Systems with Wind Generation. G. Strbac et al., January.

—— (2007b). Integration of Distributed Generation into the UK Power System Summary Report.

DTI / DETR / Ofgem (2001). Embedded Generation Working Group. Available from www.dti.gov.uk/energy.

DTI / Ofgem (2005). Regulation of Offshore Electricity Transmission, a Joint Consultation by DTI/Ofgem. July. www.dti.gov.uk.

DTI and the Scottish Office (1995). *The Prospects for Nuclear Power in the UK*. Cm 2860. HMSO. May.

Dutch Innovation Platform (2006). The Dutch Innovation Platform. The Hague.

EA Technology (2003). Solutions for the Connection and Operation of Distributed Generation. URN 03/1195. www.dti.gov.uk/files/file15187.pdf?nourl=www.dti.gov.uk/publications/pdflink/&pubpdfdload=03%2F1195.

—— (2006). A Technical Review and Assessment of Active Network Management Infrastructures and Practices, a Report for the ENSG. URN 06/1196. www.dti.gov.uk/files/file30559.pdf?nourl=www.dti.gov.uk/publications/pdflink/&pubpdfdload=06%2F1196.

ECN (2004). Dutch energy policies from a European perspective. Major developments in 2003. ECN-P-04-001. Petten, ECN: 1–68.

Econnect (2006a). Assessing the Feasibility of Establishing Registered Power Zones on Northern/Yorkshire Electricity Network. DTI URN 06/1084. www.ensg.gov.uk/assets/final_report_kel003330000.pdf.

—— (2006b). Accommodating Distributed Generation, a Report to the DTI. URN 06/1571. www.dti.gov.uk/files/file31648.pdf.

EDF (2006). Energy Review Submission. April. www.edfenergy.com/core/energyreview/edfenergy-energy_review_response_main_document_v4-3.pdf.

Ekins, P. (2000). *Economic Growth and Environmental Sustainability: The Prospects for Green Growth* (London: Routledge).

—— (2006). Time to get real. *Inside Track*, 15 (Winter): 8–11.

Elliott, D. (2007a). Supporting renewables: feed in tariffs and quota/trading systems. In D. Elliott (ed.), *Sustainable Energy* (Houndmills, Basingstoke: Palgrave Macmillan).

—— (2007b). Open University, Block 4, Innovation: Designing for a Sustainable Future, prepared for the course team by David Elliott.
Elzen, B. and A. Wieczorek (2005). Transitions towards sustainability through system innovation. *Technological Forecasting and Social Change*, 72: 651–61.
Energy Act (2004). www.opsi.gov.uk/ACTS/acts2004/20040020.htm.
Energy Networks Association (2005). G85 Innovation in Electrical Distribution Network Systems: a Good Practice Guide, Issue 1. www.energynetworks.org/spring/engineering/cms01/CMDocuments/contentManDoc_75_10f5bd06-d7ff-45a2-88e5-939ac1a49aec.pdf.
—— (2006). The State of Our Networks: Electricity and Gas in the UK 2006–2050. www.energynetworks.org/spring/mediacentre/cms03/CMDocuments/contentManDoc_56_ea7207e7-05d7-46f5-b106-6d21ed10e699.pdf.
Environmental Change Institute (2005). The 40% House. Available from www.eci.ox.ac.uk.
—— (2006). Predict and Decide. Available from www.eci.ox.ac.uk.
European Commission (EC) (2005a). The Support of Electricity from Renewable Energy Sources. COM (2005) 627 Final. http://europa.eu.int/comm/energy/res/biomass_action_plan/doc/2005_12_07_comm_biomass_electricity_en.pdf.
—— (2005b). Concerted Action for Offshore Wind Energy Development. Work Package 8: Grid Issues. www.offshorewindenergy.org/cod.
—— (2006). Fuelling Our Future: The EC sets out its vision for an Energy Strategy for Europe. http://europa.eu/rapid/pressReleasesAction.do?reference=IP/06/282.
Food Standards Agency (2005). Radioactivity in Food and the Environment 2004. www.food.gov.uk/science/surveillance/radiosurv/rife10.
Foxon, T. (2003). Inducing Innovation for a Low Carbon Future: Drivers, Barriers and Policies, a Report for the Carbon Trust. www.carbontrust.co.uk/Publicsites/cScape.CT.PublicationsOrdering/PublicationAudit.aspx?id=CT-2003-07.
Foxon, T., R. Gross et al. (2005). UK innovation systems for new and renewable energy technologies: drivers, barriers and system failures. *Energy Policy*, 33: 2123–37.
Foxon, T. and P. Pearson (forthcoming). Towards improved policy processes for promoting innovation in renewable electricity technologies in the UK. *Energy Policy*.
Geels, F. (2004a). From sectoral systems of innovation to socio-technical systems: insights about dynamics and change from sociology and institutional theory. *Research Policy*, 33(6–7): 897–920.
—— (2004b). Understanding system innovations: a critical literature review and a conceptual synthesis. In B. Elzen, F. Geels and K. Green (eds), *System Innovation and the Transition to Sustainability* (Cheltenham, Northampton: Edward Elgar).
—— (2005a). Processes and patterns in transitions and system innovations: refining the co-evolutionary multi-level perspective. *Technological Forecasting and Social Change*, 72(6): 681–96.
—— (2005b). Co-evolution of technology and society: the transition in water supply and personal hygiene in the Netherlands (1850–1930) – a case study in multi-level perspective. *Technology in Society*, 27(3): 363–97.
—— (2005c). The dynamics of transitions in socio-technical systems: a multi-level analysis of the transition pathway from horse-drawn carriages to automobiles (1860–1930). *Technology Analysis and Strategic Management*, 17(4): 445–76.

—— (2006). Co-evolutionary and multi-level dynamics in transitions: the transformation of aviation systems and the shift from propeller to turbojet (1930–1970). *Technovation*, 26(9): 999–1016.

Grayling, T. et al. (2005). Climate Commitment – Meeting the UK's 2010 CO_2 emissions target. IPPR.

Greenpeace International (2007). *The Economics of Nuclear Power.*

Grubb, M. (2006). Climate Change Solutions in Theory and Practice: national and international dimensions. IPS Conference on Climate Change: the Policy Dimension. Wellington, New Zealand, 6 October.

Hannah, L. (1979). *Electricity Before Nationalisation: a Study of the Development of the Electricity Supply Industry in Britain to 1948* (London: Macmillan).

Harremoks, P. et al. (2002). *The Precautionary Principle in the Twentieth Century: Late Lessons from Early Warnings* (London: Earthscan).

Health and Safety Executive (2006). HSE Review of the Pre-Licensing Process for Potential New Build of Nuclear Power Stations. www.hse.gov.uk/consult/condocs/energyreview/discussion.htm.

Helm, D. (2004). *Energy, the State and the Market: British Energy Policy Since 1979* (Oxford: Oxford University Press).

—— (2005). The assessment: the new energy paradigm. *Oxford Review of Economic Policy*, 21(1): 1–18.

Hellsmark, H. Incumbents versus new entrants in the exploitation of renewable energy technologies. Available from hans.hellsmark@mot.chalmers.se.

Hill, M. (ed.) (1997). *The Policy Process: A Reader* (London: Prentice Hall/Harvester Wheatsheaf).

Hood, C. et al. (1999). *Regulation Inside Government: Waste-Watchers, Quality Policy and Sleaze-Busters* (Oxford: Oxford University Press).

Hoogma, R., R. Kemp, J. Schot and B. Truffer (2002). *Experimenting for Sustainable Transport: The Approach of Strategic Niche Management* (London: Spon Press).

House of Commons Environmental Audit Committee (2002). A Sustainable Energy Strategy? Renewables and the PIU Review. Fifth Report of Session 2001–02.

—— (2006). Keeping the Lights On: Nuclear, Renewables and Climate Change. Sixth Report of Session 2005–06. HC-584-1. www.publications.parliament.uk/pa/cm200506/cmselect/cmenvaud/584/584i.pdf.

House of Commons Science and Technology Committee (2003). Towards a Non-Carbon Fuel Economy: Research, Development and Demonstration. Fourth Report of Session 2002–03. HC 55-1.

House of Commons Trade and Industry Committee (2006). The Work of the NDA and UKAEA. Sixth Report of Session 2005–06. HC 1028. www.publications.parliament.uk/pa/cm200506/cmselect/cmtrdind/1028/1028.pdf.

House of Lords Science and Technology Select Committee (1992). *The Waldegrave Report.*

—— (1999). Management of Nuclear Waste. Third Report Session 1998–99. www.publications.parliament.uk/pa/ld199899/ldselect/ldsctech/41/4101.htm; www.psiru.org/reports/2005-09-E-Nuclear.pdf.

Hughes, T.P. (1983). *Networks of Power: Electrification in Western Society 1880–1930* (London: Johns Hopkins University Press).

—— (1987). The evolution of large technical systems. In W. Bijker, T. Hughes and T. Pinch (eds), *The Social Construction of Technological Systems: New Directions in the Sociology and History of Technology* (London: MIT Press).

Ilex Energy Consulting (2002). NETA – The Next Phase. www.ilexenergy.com/pages/NetaTheNextPhase26Mar02.pdf.

International Energy Agency (2004). Energy Policies of IEA Countries: Finland 2003 Review. www.iea.org/textbase/nppdf/free/2000/finland2003.pdf.

Intergovernmental Panel on Climate Change (IPCC) (2007), Summary for Policymakers of Volume 1 of Climate Change 2007, The Fourth Assessment Report (AR4), 2 February.

Islas, J. (1999). The gas turbine: a new technological paradigm in electricity generation. *Technological Forecasting and Social Change*, 60: 128–48.

Jacobsson, S. and A. Bergek (2002). Energy System Transformation: the evolution of technological systems in renewable energy technology. Available from stajac@mot.chalmers.se or anneb@eki.liu.se.

—— (2004). Transforming the energy sector: the evolution of technological systems in renewable energy technology. *Industrial and Corporate Change*, 13(5): 815–49.

Jänicke, M. (2004). Industrial Transformation Between Ecological Modernisation and Structural Change. Governance for Industrial Transformation. Proceedings of the 2003 Conference on the Human Dimensions of Global Environmental Change. K. Jacob, M. Binder and A. Wieczorek. Berlin, Environmental Policy Research Centre: 201–7.

Joskow, P. (2006). Incentive Regulation in Theory and Practice: Electricity Distribution and Transmission Networks. MIT. 21 January. Prepared for the National Bureau of Economic Research Conference on Economic Regulation, 9/10/2005.

Kemp, R. and D. Loorbach (2004). Dutch policies to manage the transition to sustainable energy. In R. Kemp and F. Beckenbach et al. (eds), *Jahrbuch Ökologische Ökonomik: Innovationen und Transformation*. Band 4, Metropolis, Marburg.

—— (2005). Dutch Policies to Manage the Transition to Sustainable Energy. *Jahrbuch Ökologische Ökonomik: Innovationen und Transformation*. F. Beckenbach, U. Hampicke, C. Leipertet al. Marburg, Metropolis Verlag. 4: 123–50.

Kemp, R. and J. Rotmans (2004). Managing the transition to sustainable mobility. In B. Elzen, F.W. Geels and K. Green (eds), *System Innovation and the Transition to Sustainability: Theory, Evidence and Policy* (Cheltenham: Edward Elgar).

Kemp, R., A. Rip and J. Schot (2001). Constructing transition paths through the management of niches. In R. Garud and P. Karnoe (eds), *Path Dependence and Creation* (Mahwah, NJ: Lawrence Erlbaum Associates).

Kemp, R., J. Schot and R. Hoogma (1998). Regime shifts to sustainability through processes of niche formation: the approach of Strategic Niche Management. *Technology Analysis and Strategic Management*, 10(2): 175–95.

Kern, F. and A. Smith (2007, forthcoming). Restructuring energy systems for sustainability? Energy transition policy in the Netherlands. Sussex Energy Group, SPRU, University of Sussex.

King, D. (2005). Chief Scientist backs nuclear power revival. Interview with the *Guardian* newspaper, 21 October. www.guardian.co.uk/science/2005/oct/21/energy.greenpolitics.

Klein, A. et al. (2006). Evaluation of Different Feed-in Tariff Design Options – best practice paper for the International Feed-in Co-operation. Available from Fraunhofer Institute for Systems and Innovation Research.

Kline, S. and N. Rosenberg (1986). An overview of innovation. In R. Landau and N. Rosenberg (eds), *The Positive Sum States* (Washington, DC: National Academy Press).

Lamb, A. (2007). Sustainable Energy Service Companies: analysis of a low carbon innovation niche in the UK. MBA thesis. Warwick Business School.

Lash, J. and F. Wellinton (2007). Competitive advantage on a warming planet. *Harvard Business Review*, 95.

Lashof, D. and D. Ahuja (1990). Relative contributions of greenhouse gas emissions to global warming. *Nature*, 344: 529–31.

Loorbach, D. and J. Rotmans (2006). Managing transitions for sustainable development. In X. Olsthoorn and A.J. Wieczorek (eds), *Understanding Industrial Transformation: Views from Different Disciplines* (Dordrecht: Springer).

Loughhead, J. (2005). Interview on Newsnight, BBC2, 21 November.

Mackerron, G. (1996). Nuclear power under review. In J. Surrey (ed.), *The British Electricity Experiment, Privatisation: the Record, the Issues, the Lessons* (London: Earthscan).

MAF (2006). Sustainable Land Management and Climate Change: Options for a Plan of Action. Available from the NZ Ministry of Agriculture and Forestry website: www.maf.govt.nz/climatechange.

Marr, A. (2007). *A History of Modern Britain* (London: Macmillan).

Meadowcroft, J. (2005). Environmental political economy, technological transitions and the state. *New Political Economy*, 10(4): 479–98.

MED (2006). Transitional Measures: Options to Move Towards a Low Emissions Electricity and Stationary Energy Supply and to Facilitate a Transition to Greenhouse Gas Pricing in the Future. Available from the NZ Ministry of Economic Development website: www.med.govt.nz.

Meyer, N. and A.L. Koefoed (2003). Danish energy reform: policy implications for renewables. *Energy Policy*, 31: 597–607.

Ministry for the Environment, Nature Conservation and Nuclear Safety, Germany. Evaluation of different feed-in tariff design options – best practice paper. Available from www.eneuerbare-energien.de.

Mitchell, C. (1995). The Renewable NFFO – a review. *Energy Policy*, 23(12): 1077–91.

—— (2000a). The Non-Fossil Fuel Obligation and its future. *Annual Review of Energy and Environment*, 25: 285–312.

—— (2000b). Neutral regulation – the vital ingredient for a sustainable energy future. In C. Mitchell (ed.), *Renewable Energy – Issues for the New Millennium. Energy and Environment*, Special Issue, 11(4): 377–90, October.

—— (ed.) (2004). Special edition: energy policy for a sustainable future. *Energy Policy*, 32(17): 1887–9.

Mitchell, C. and P. Connor (2002). Review of Current Electricity Policy and Regulation, UK Case Study. Report for the Sustelnet Project.

—— (2004). Renewable energy policy in the UK 1990–2003. *Energy Policy*, 32(17): 1935–47.

Mitchell, C. and A. White (1999). *Regulation of Distribution and Supply, The Energy Report*, DTI, The Stationery Office, London, UK.

Mitchell, C. and B. Woodman (2004). *The Burning Question – Is the UK on Course for a Low Carbon Economy?* (IPPR Press).

—— (2006). New nuclear power: implications for a sustainable energy system. A Warwick Business School and Green Alliance report. Available from www.green-alliance.org.uk/publications/NewNuclearPowerRpt/.

Mitchell, C., D. Bauknecht and P. Connor (2006). Effectiveness through risk reduction: a comparison of the renewable obligation in England and Wales and the feed in system in Germany. *Energy Policy*, 34: 297–305.

Moran, M. (2003). *The British Regulatory State* (Oxford: Oxford University Press).

Mott MacDonald and British Power International (BPI) (2004a). DG-BPQ Analysis, Summary of Findings. Final Report. March.

—— (2004b). Innovation in Electricity Distribution Networks. Final Report.

National Audit Office (2002). Pipes and Wires. Report by the Comptroller and Auditor General. HC 723. Session 2001–2002.

—— (2005). Renewable Energy. HC 210. Session 2004–05, 11 February.

Nelson, R. and S. Winter (1977). In search of useful theory of innovation. *Research Policy*, 6: 36–76.

New Statesman (2007). *The Future of Energy*. Special supplement, 2 July, pp. 29–30.

NZES (2006). Powering Our Future: Towards a Sustainable Low Emissions Energy System – Draft New Zealand Energy Strategy to 2050. Available from www.med.govt.nz/energy.

NZEECS (2006). New Zealand Energy Efficiency and Conservation Strategy. www.eeca.govt.nz/eeca-library/eeca-reports/neecs/report/draft-nzeecs-06.pdf.

North, D. (1990). *Institutions, Institutional Change and Economic Performance* (Cambridge: Cambridge University Press).

Nuclear Decommissioning Authority (NDA) (2006). Approved Strategy for Clean-up of UK's Nuclear Sites Published. Press release, 30 March. www.nda.gov.uk/documents/news_release_-_national.pdf.

Official Journal of the European Union (2003). State aid – United Kingdom – restructuring aid in favour of British Energy plc. OJ C, 180: 5–28. http://eur-lex.europa.eu/LexUriServ/LexUriServ.do?uri=OJ:C:2003:180:0005:0028:EN:PDF.

Ofgem (2001). Review of the New Electricity Trading Arrangements (NETA) and the Impact on Smaller Generators. Available from www.ofgem.gov.uk/Markets/WhlMkts/CustandIndustry/DemSideWG/Archive/393-31aug01_pub.pdf.

—— (2002). Report to the DTI on the Review of the Initial Impact of NETA on Smaller Generators. 21 February. Available from www.ofgem.gov.uk/Markets/WhlMkts/Archive/111-31aug01.pdf.

—— (2003). Innovation and Registered Power Zones. Discussion Paper. July.

—— (2004a). Electricity Distribution Price Control Review. Regulatory Impact Assessment for Distributed Generation and the Structure of Distribution Charges. March.

—— (2004b). Electricity Distribution Price Control Review, Final Proposals. 265/04. November. www.ofgem.gov.uk/temp/ofgem/cache/cmsattach/9416_26504.pdf?wtfrom=/ofgem/work/index.jsp§ion=/areasofwork/distpricecontrol.

—— (2004c). Electricity Distribution Price Control Review. Impact Assessment. 265b/04. November. www.ofgem.gov.uk/temp/ofgem/cache/cmsattach/9565_26504b.pdf?wtfrom=/ofgem/work/index.jsp§ion=/areasofwork/distpricecontrol.

—— (2004d). Offshore Electricity Transmission. Letter from Andrew Walker, 30 December 2004. www.ofgem.gov.uk/Networks/Trans/Offshore/Consultation-DecisionsResponses/Documents1/9307-offshore_letter.pdf.

—— (2005). Further Details of the RPZ Scheme. Guidance Document. Version 1, April.

—— (2006a). Renewables Obligation: Third Annual Report, 35/06, February.

—— (2006b). Sustainable Development Report 2006. Available from www.ofgem.gov.uk/Sustainability/Pages/Sustain.aspx.

—— (2006c). Transmission Price Control Review: Final Proposals, 206/06. www.ofgem.gov.uk/temp/ofgem/cache/cmsattach/17916_20061201_TPCR_Final_Proposals_in_v71_6_Final.pdf?wtfrom=/ofgem/work/index.jsp§ion=/areasofwork/transpcr.

—— (2006d). Decision in relation to the introduction of Engineering Recommendation P2/6 and the consequential Distribution Code amendments. June. www.ofgem.gov.uk/temp/ofgem/cache/cmsattach/15428_9906.pdf?wtfrom=/ofgem/whats-new/archive.jsp.

—— (2006e). Letter to Bernard Bulkin, Sustainable Development Commission. 2 November. www.ofgem.gov.uk/temp/ofgem/cache/cmsattach/17519_bulkin.pdf.

—— (2006f). Open Letter Consultation on the IFI and RPZ Schemes for Distribution Network Operators. www.ofgem.gov.uk/temp/ofgem/cache/cmsattach/16958_181_06.pdf?wtfrom=/ofgem/work/index.jsp§ion=/areasofwork/ifirpz.

—— (2007a). Renewables Obligation: Annual Report 2005–06. 35/06.

—— (2007b). Open Letter Consultation on the IFI and RPZ Schemes for Distribution Network Operators. www.ofgem.gov.uk/temp/ofgem/cache/cmsattach/18783_2507.pdf?wtfrom=/ofgem/work/index.jsp§ion=/areasofwork/ifirpz.

—— (2007c). Offshore Electricity Transmission – Second Scoping Document, 58/07. March. www.ofgem.gov.uk.

Owen, G. (2006). Sustainable development duties: new roles for economic regulators. *Utilities Policy*, 14: 208–17.

Parliamentary Office of Science and Technology (1997). Radioactive Waste – Where Next? POST Report Summary. www.parliament.uk/post/pn106.pdf.

Performance and Innovation Unit (PIU) (2002). The Energy Review. www.strategy.gov.uk/downloads/su/energy/TheEnergyReview.pdf.

Pierson, P. (2000). Increasing returns, path dependence and the study of politics. *American Political Science Review*, 94(2): 251–67.

Ragwitz, M. and C. Huber (2006). Feed-in Systems in Germany and Spain – a comparison. Available from Fraunhofer Institute for Systems and Innovation Research.

Raven, R. (2006). Towards alternative trajectories? Reconfigurations in the Dutch electricity regime. *Research Policy*, 35: 581–95.

Renewable Energy Association (REA) (2007a). Energy Review or Smoke-screen? Annual conference. Philip Wolfe. Available from www.r-e-a.net/article_default_view.fcm?articleid=1619.

—— (2007b). Evidence to the Select Committee on Science and Technology Inquiry – Renewable Energy Generation Technologies. Available from www.r-e-a.net.

—— (2007c). Europe's 20% Renewables Target – the UK Contribution. Available from www.r-e-a.net.

REN21 (2007). Renewable Energy Policy Network for the 21st Century. Available from www.ren21.net.

Rickerson, W. and R. Crace (2007). The Debate over Fixed Price Incentives for Renewable Electricity in Europe and the United States: Fallout and Future Directions. The Heinrich Boll Foundation, February.

Rip, A. and R. Kemp (1998). Technological change. In S. Rayner and E. Malone (eds), *Human Choices and Climate Change*, Vol. 2 (Columbus, OH: Batelle Press).

Rosenberg, N. (1982). *Inside the Black Box: Technology and Economics* (Cambridge: Cambridge University Press).

Rotmans, J. (2005). *Societal Innovation: Between Dream and Reality Lies Complexity* (Rotterdam: Erasmus University).

Rotmans, J., R. Kemp and M. van Asselt (2001). More evolution than revolution: transition management in public policy. *Foresight*, 3(1): 15–31. www.icis.unimaas.nl/publ/downs/01_12.pdf.

Rowell, A. (2006). Plugging the Gap. *Guardian*, 3 May. http://politics.guardian.co.uk/foi/story/0,,1770183,00.html.

Royal Commission on Environmental Pollution (RCEP) (1976) Sixth Report, Nuclear Power and the Environment. Cmnd 6618.

—— (2000). Twenty Second Report, Energy – The Changing Climate. Cm 4749. www.rcep.org.uk/newenergy.htm.

Saviotti, P.P. (1986). Systems theory and technological change. Futures (December): 773–85.

Scott, J. (1998). *Seeing Like a State: How Certain Schemes to Improve the Human Condition Have Failed* (New Haven, CT: Yale University Press).

Scottish Executive (2006). Statutory Consultation on Renewables Obligation (Scotland). Available from www.scotland.gov.uk.

Shove, E. (2003). *Comfort, Cleanliness and Convenience* (Oxford: Berg).

Shove, E. and G. Walker (2007). CAUTION! Transitions ahead: politics, practice, and sustainable transition management. *Environment and Planning* A, 39: 763–70.

Skidmore, P. et al. (2004). *The Long Game* (London: Demos).

Smith, A. (2006). Niche-based approaches to sustainable development: radical activists versus strategic managers in Reflexive Governance of sustainable socio-technical transitions. *Research Policy*, 34: 1491–510.

Smith, A. and A. Stirling (2006). *Inside or Out? Open or Closed? Positioning the Governance of Sustainable Technology*. SPRU Electronic Working Paper Series No. 148. Brighton, SPRU.

Smith, A., F. Berkhout and A. Stirling (2004). Socio-technical regimes and transition contexts. In B. Elzen, F. Geels and K. Green (eds), *System Innovation and the Transition to Sustainability: Theory, Evidence and Policy* (Camberley: Edward Elgar Publishing).

SPRU and NERA Economic Consulting (2006). The Economics of Nuclear Power: An evidence-based report for the Sustainable Development Commission. www.sd-commission.org.uk/publications/downloads/Nuclear-paper4-Economics.pdf.

Stenzel, T. and A. Frenzel (2007). Mutiny Against the Bounty – how firms shape subsidy regulation and new technologies in European Electricity Markets. Academy of Management Conference, Philadelphia, USA, 3–8 August.

Stern, N. (2006). Stern Review: The Economics of Climate Change. www.hm-treasury.gov.uk/independent_reviews/stern_review_economics_climate_change/stern_review_report.cfm.
Stirling A. (2005). Opening up or closing down: analysis, participation and power in the social appraisal of technology. In M. Leach et al. (eds), *Science and Citizens* (London: Zed).
—— (2006a). Deliberate Futures: Precaution and Progress in Social Choice in Sustainable Technology. Available from www.sussex.ac.uk/spru.
—— (2006b). Uncertainty, precaution and Sustainability: towards more reflexive governance for sustainability. In J-P. Voss and R. Kemp (eds), *Sustainability and Reflexive Governance* (Cheltenham: Edward Elgar).
Stockholm Environment Institute (2005). Social Cost of Carbon: A Closer Look at Uncertainty. www.defra.gov.uk/environment/climatechange/research/carboncost/pdf/sei-scc-report.pdf.
Strbac, G. and N. Jenkins (2001). PIU Working Paper on Network Security of the Future Electricity System. www.cabinetoffice.gov.uk/strategy/work_areas/energy/background.asp.
Surrey, J. (ed.) (1996). *The British Electricity Experiment – Privatisation: The Record, the Issues, the Lessons* (London: Earthscan).
Sustainable Development Commission (SDC) (2007). Lost in Transmission: The Role of Ofgem in a Changing Climate. Available from www.sd-commission.org.uk/publications/downloads/SDC_ofgem_report.pdf.
Sustelnet (2005). Policy and Network Regulation for the Integration of Distribution Generation and Renewables in Energy Supply. www.ecn.nl/docs/library/report/2005/rx05173.pdf.
Szarka, J. (2006). Wind power, policy learning and paradigm change. *Energy Policy*, 34: 3041–8.
Szarka, J. and I. Bluhdorn (2006). *Wind Power in Britain and Germany: Explaining contrasting development paths*. Anglo-German Foundation for the Study of Industrial Society.
Thomas, S. (2002). The Economics of New Nuclear Power Plants and Electricity Liberalisation: Lessons for Finland from British Experience. www.psiru.org/reports/2002-01-E-Finnuclear.doc.
—— (2005). The Economics of Nuclear Power: Analysis of Recent Studies. www.psiru.org/reports/2005-09-E-Nuclear.pdf.
Transport- og Energiministeriet (2007). A Visionary Danish Energy Policy 2025. January.
TVO (2006). Olkiluoto 3 – Current News in August. www.tvo.fi/930.htm.
Twomey, P. and K. Neuhoff (2005). Market Power and Technological Bias – the case of electricity generation. August 2005. CWPE 0532 and EPRG 01.
UCL Environment Institute (2007). UK Greenhouse Gas Emissions – are we on target?
UKERC (2006). The Costs and Impacts of Intermittency: an assessment of the evidence on the costs and impacts of intermittent generation on the British electricity system. Available from www.ukerc.ac.uk.
—— (2007). Investment in Electricity Generation – the role of costs, incentives and risks. May 2007. Available from www.ukerc.ac.uk.
Unruh, G.C. (2002). Escaping carbon lock-in. *Energy Policy*, 30(4): 317–25.
Utilities Act (2000). www.opsi.gov.uk/acts/acts2000/20000027.htm.

van Rooijen, S.N.M. and M.T. van Wees (2006). Green electricity policies in the Netherlands: an analysis of policy decisions. *Energy Policy*, 34(1): 60–71.

Verbong, G. and F. Geels (2007). The ongoing energy transition: lessons from a socio-technical, multi-level analysis of the Dutch electricity system (1960–2004). *Energy Policy*, 35(2): 1025–37.

Vickers, J. and G. Yarrow (1988). *Privatisation: An Economic Analysis* (Cambridge, MA: MIT Press).

VROM (2003). Transition Progress Report. Making Strides towards Sustainability, Directorate-General for the Environment: 1–20.

—— (2006). Future Environment Agenda: clean, clever, competitive. The Hague.

VROM-Raad and AER (2004). Energy Transition: A Climate for New Opportunities.

Walker, W. (2000). Entrapment in large technology systems: institutional commitment and power relations. *Research Policy*, 29(7–8): 833–46.

Wicks, M. (2006). Reported in the *Guardian* newspaper, 11 October. No subsidies for nuclear, says Energy Minister. http://business.guardian.co.uk/story/0,,1892310,00.html.

Williamson, O. (2000). The new institutional economics: taking stock, looking ahead. *Journal of Economic Literature*, 38(3): 595–613.

Winner, L. (1977). *Autonomous Technology: Technics-Out-of-Control as a Theme in Political Thought* (Cambridge, MA: MIT Press).

Wiser, R., K. Porter and M. Bolinger (2006). Comparing State Portfolio Standards and System Benefits Charges Under Restructuring. Lawrence Berkeley National Laboratory. Available from http://eetd.lbl.gov/EA/EMP/re-pubs.html.

Woodman, B. (2007a, forthcoming). Innovation in Distribution Networks. UKERC Working Paper.

—— (2007b, forthcoming). Rewriting history: a new future for nuclear power in the UK.

—— (2007c, forthcoming). Offshore Transmission Networks – how to regulate slowly. UKERC Working Paper.

World Future Council (2007). Feed-in Tariffs – Energy of our Future: A guide to one of the world's best environmental policies.

WWF (2005). PowerSwitch! Scenarios for a Clean Future. 28 November. www.panda.org/about_wwf/what_we_do/climate_change/publications/energy_efficiency_publications.cfm?uNewsID=52140.

Index

Compiled by Sue Carlton